Exploring Water Resources

GIS Investigations for the Earth Sciences

Michelle K. Hall-Wallace
C. Scott Walker
Larry P. Kendall
Christian J. Schaller
The University of Arizona

BROOKS/COLE

™

THOMSON LEARNING

Australia • Canada • Mexico • Singapore • Spain • United Kingdom • United States

BROOKS/COLE

™

THOMSON LEARNING

Editor: *Keith Dodson*
Assistant Editor: *Carol Ann Benedict*
Editorial Assistant: *Helen Kranz*
Marketing Manager: *Ann Caven*
Advertising Project Manager: *Laura Hubrich*
Project Manager, Editorial Production: *Tom Novack*

Print/Media Buyer: *Kristine Waller*
Permissions Editor: *Sue Ewing*
Cover Designer: *Denise Davidson*
Cover Image: *Digital Vision*
Cover Printer: *C&C Offset Printing Co., Ltd*
Printer: *C&C Offset Printing Co., Ltd*

For more information about our products, contact us at:
Thomson Learning Academic Resource Center
1-800-423-0563

For permission to use material from this text, contact us by:
Phone: 1-800-730-2214   Fax: 1-800-730-2215
Web: http://www.thomsonrights.com

All products used herein are used for identification
purposes only and may be trademarks or registered
trademarks of their respective owners.

Maps and screen shots include data from ESRI Data
and Maps. Data Copyright © ESRI 2002.

ESRI and ArcView are registered trademarks in the
United States and are either trademarks or registered
trademarks in all other countries in which they
are used. The ArcView logo is a trademark of
Environmental Systems Research Institute, Inc.

Earth surface image data used in the Earth globe
illustration on page 47 is © 2001, TerraMetrics, Inc.
Used with permission.

Development of these materials was supported, in
part, by the National Science Foundation under
Grant No. DUE-9555205. Any opinions, findings,
and conclusions or recommendations expressed
in these materials are those of the authors and do
not necessarily reflect the views of the National Science
Foundation.

Library of Congress Control Number: 2002102318

ISBN: 0-534-39156-7

Brooks/Cole—Thomson Learning
511 Forest Lodge Road
Pacific Grove, CA 93950
USA

Asia
Thomson Learning
5 Shenton Way #01-01
UIC Building
Singapore 068808

Australia
Nelson Thomson Learning
102 Dodds Street
South Melbourne, Victoria 3205
Australia

Canada
Nelson Thomson Learning
1120 Birchmount Road
Toronto, Ontario M1K 5G4
Canada

Europe/Middle East/Africa
Thomson Learning
High Holborn House
50/51 Bedford Row
London WC1R 4LR
United Kingdom

Latin America
Thomson Learning
Seneca, 53
Colonia Polanco
11560 Mexico D.F.
Mexico

Spain
Paraninfo Thomson Learning
Calle/Magallanes, 25
28015 Madrid, Spain

Acknowledgments

The authors wish to thank the many students, teachers, and scientists who, through their use of these materials, provided critical reviews and helped us develop insight into how GIS can be most effectively used as a learning and teaching tool.

A significant number of people contributed directly or indirectly to the development of this module, but a few are especially notable. Particular thanks go to Terry Wallace, Joshua Hall, Christine Donovan, and Tekla Cook who tested the investigations in their classrooms and provided valuable feedback that helped us to improve the content and design of the activities. Jennifer Weeks contributed significantly to the final rounds of editing and revision, while Terry Wallace, Joseph Watkins, and James Washburne provided content reviews. We also appreciate the considerable efforts of our student assistants Marie Renwald, Anne Kramer Huth, Tammy Baldwin, Sara McNamara, Megan Sayles, and Ted Stude.

We are indebted to the numerous scientists who took the time to learn about our project and share critical research data or expertise that added greatly to the quality of the investigations. Special thanks go to Don Pool of the USGS Tucson field office and to the Pima County Department of Transportation Technical Services Division for their assistance in obtaining data used in Unit 4. Finally, we are grateful to the agencies, publishers, and individuals that have given us permission to include their outstanding illustrations, stories, and photos.

The SAGUARO Project
Michelle K. Hall-Wallace, Director
Department of Geosciences
The University of Arizona
1040 E Fourth Street • Tucson, AZ 85721-0077
saguaro@geo.arizona.edu

Science And GIS Unlocking Analysis & Research Opportunities
http://saguaro.geo.arizona.edu

ESRI Software License Agreement

This is a license agreement and not an agreement for sale. This license agreement (Agreement) is between the end user (Licensee) and Environmental Systems Research Institute, Inc. (ESRI), and gives Licensee certain limited rights to use the proprietary ESRI® desktop software and software updates, sample data, online and/or hard-copy documentation and user guides, including updates thereto, and software keycode or hardware key, as applicable (hereinafter referred to as "Software, Data, and Related Materials"). All rights not specifically granted in this Agreement are reserved to ESRI.

Reservation of Ownership and Grant of License: ESRI and its third party licensor(s) retain exclusive rights, title, and ownership of the copy of the Software, Data, and Related Materials licensed under this Agreement and, hereby, grants to Licensee a personal, nonexclusive, nontransferable license to use the Software, Data, and Related Materials based on the terms and conditions of this Agreement. From the date of receipt, Licensee agrees to use reasonable effort to protect the Software, Data, and Related Materials from unauthorized use, reproduction, distribution, or publication.

Copyright: The Software, Data, and Related Materials are owned by ESRI and its third party licensor(s) and are protected by United States copyright laws and applicable international laws, treaties, and/or conventions. Licensee agrees not to export the Software, Data, and Related Materials into a country that does not have copyright laws that will protect ESRI's proprietary rights. Licensee may claim copyright ownership in the Simple Macro Language (SML™) macros, AutoLISP® scripts, AtlasWare™ scripts, and/or Avenue™ scripts developed by Licensee using the respective macro and/or scripting language.

Permitted Uses:
- Licensee may use the number of copies of the Software, Data, and Related Materials for which license fees have been paid on the computer system(s) and/or specific computer network(s) for Licensee's own internal use. Licensee may use the Software, Data, and Related Materials as a map/data server engine in an Internet and/or Intranet distributed computing network or environment provided the appropriate, additional license fees are paid. If the Software, Data, and Related Materials contain dual media (i.e., both 3.5-inch diskettes and CD–ROM), then Licensee may only use one (1) set of the dual media provided. Licensee may not use the other media on another computer system(s) and/or specific computer network(s), or loan, rent, lease, or transfer the other media to another user.
- Licensee may install the number of copies of the Software, Data, and Related Materials for which license or update fees have been paid onto the permanent storage device(s) on the computer system(s) and/or specific computer network(s).
- Licensee may make routine computer backups but only one (1) copy of the Software, Data, and Related Materials for archival purposes during the term of this Agreement unless the right to make additional copies is granted to Licensee in writing by ESRI.
- Licensee may use, copy, alter, modify, merge, reproduce, and/or create derivative works of the online documentation for Licensee's own internal use. The portions of the online documentation merged with other software, hard copy, and/or digital materials shall continue to be subject to the terms and conditions of this Agreement and shall provide the following copyright attribution notice acknowledging ESRI's proprietary rights in the online documentation: "Portions of this document include intellectual property of ESRI and are used herein by permission. Copyright © 200_ Environmental Systems Research Institute, Inc. All Rights Reserved."
- Licensee may use the Data that are provided under license from ESRI and its third party licensor(s) as described in the Distribution Rights section of the online Data Help files.

Uses Not Permitted:
- Licensee shall not sell, rent, lease, sublicense, lend, assign, time-share, or transfer, in whole or in part, or provide unlicensed third parties access to prior or present versions of the Software, Data, and Related Materials, any updates, or Licensee's rights under this Agreement.
- Licensee shall not reverse engineer, decompile, or disassemble the Software, or make any attempt to unlock or bypass the software keycode and/or hardware key used, as applicable, subject to local law.
- Licensee shall not make additional copies of the Software, Data, and/or Related Materials beyond that described in the Permitted Uses section above.
- Licensee shall not remove or obscure any ESRI copyright or trademark notices.
- Licensee shall not use this software for more than one hundred twenty (120) days from the date that the software was installed. At the end of this period, users must remove the time limited software from their computers or purchase fully licensed software. Students and instructors in the United States may purchase fully licensed individual copies of the software from ESRI telesales at 1-800-GIS-XPRT.

Term: The license granted by this Agreement shall commence upon Licensee's receipt of the Software, Data, and Related Materials and shall continue until such time that (1) Licensee elects to discontinue use of the Software, Data, and Related Materials and terminates this Agreement or (2) ESRI terminates for Licensee's material breach of this Agreement. Upon termination of this Agreement in either instance, Licensee shall return to ESRI the Software, Data, Related Materials, and any whole or partial copies, codes, modifications, and merged portions in any form. The parties hereby agree that all provisions, which operate to protect the rights of ESRI, shall remain in force should breach occur.

Limited Warranty: ESRI warrants that the media upon which the Software, Data, and Related Materials are provided will be free from defects in materials and workmanship under normal use and service for a period of sixty (60) days from the date of receipt. The Data herein have been obtained from sources believed to be reliable, but its accuracy and completeness, and the opinions based thereon, are not guaranteed. Every effort has been made to provide accurate Data in this package. The Licensee acknowledges that the Data may contain some nonconformities, defects, errors, and/or omissions. ESRI and third party licensor(s) do not warrant that the Data will meet Licensee's needs or expectations, that the use of the Data will be uninterrupted, or that all nonconformities can or will be corrected. ESRI and the respective third party licensor(s) are not inviting reliance on these Data, and Licensee should always verify actual map data and information. The Data contained in this package are subject to change without notice.

EXCEPT FOR THE ABOVE EXPRESS LIMITED WARRANTIES, THE SOFTWARE, DATA, AND RELATED MATERIALS CONTAINED THEREIN ARE PROVIDED "AS IS," WITHOUT WARRANTY OF ANY KIND, EITHER EXPRESS OR IMPLIED, INCLUDING, BUT NOT LIMITED TO, THE IMPLIED WARRANTIES OF MERCHANTABILITY AND FITNESS FOR A PARTICULAR PURPOSE.

Exclusive Remedy and Limitation of Liability: During the warranty period, ESRI's entire liability and Licensee's exclusive remedy shall be the return of the license fee paid for the Software, Data, and Related Materials in accordance with the ESRI Customer Assurance Program for the Software, Data, and Related Materials that do not meet ESRI's Limited Warranty and that are returned to ESRI or its dealers with a copy of Licensee's proof of payment.

ESRI shall not be liable for indirect, special, incidental, or consequential damages related to Licensee's use of the Software, Data, and Related Materials, even if ESRI is advised of the possibility of such damage.

Waivers: No failure or delay by ESRI in enforcing any right or remedy under this Agreement shall be construed as a waiver of any future or other exercise of such right or remedy by ESRI.

Order of Precedence: Any conflict and/or inconsistency between the terms of this Agreement and any FAR, DFAR, purchase order, or other terms shall be resolved in favor of the terms expressed in this Agreement, subject to the U.S. Government's minimum rights unless agreed otherwise.

Export Regulations: Licensee acknowledges that this Agreement and the performance thereof are subject to compliance with any and all applicable United States laws, regulations, or orders relating to the export of computer software or know-how relating thereto. ESRI Software, Data, and Related Materials have been determined to be Technical Data under United States export laws. Licensee agrees to comply with all laws, regulations, and orders of the United States in regard to any export of such Technical Data. Licensee agrees not to disclose or reexport any Technical Data received under this Agreement in or to any countries for which the United States Government requires an export license or other supporting documentation at the time of export or transfer, unless Licensee has obtained prior written authorization from ESRI and the U.S. Office of Export Control. The countries restricted at the time of this Agreement are Cuba, Iran, Iraq, Libya, North Korea, Serbia, and Sudan.

U.S. Government Restricted/Limited Rights: Any software, documentation, and/or data delivered hereunder is subject to the terms of the License Agreement. In no event shall the Government acquire greater than RESTRICTED/LIMITED RIGHTS. At a minimum, use, duplication, or disclosure by the Government is subject to restrictions as set forth in FAR §52.227-14 Alternates I, II, and III (JUN 1987); FAR §52.227-19 (JUN 1987) and/or FAR §12.211/12.212 (Commercial Technical Data/Computer Software); and DFARS §252.227-7015 (NOV 1995) (Technical Data) and/or DFARS §227.7202 (Computer Software), as applicable. Contractor/Manufacturer is Environmental Systems Research Institute, Inc., 380 New York Street, Redlands, CA 92373-8100, USA.

Governing Law: This Agreement is governed by the laws of the United States of America and the State of California without reference to conflict of laws principles.

Entire Agreement: The parties agree that this constitutes the sole and entire agreement of the parties as to the matter set forth herein and supersedes any previous agreements, understandings, and arrangements between the parties relating hereto and is effective, valid, and binding upon the parties.

Introduction

Getting started

Unit 1 – Global Water Reservoirs

Unit 2 – The Renewable Resource

Unit 3 – Using Water Wisely

Unit 4 – Water for a Desert City

Introduction

Thinking scientifically

An Earth scientist makes a living by observing and measuring nature. Whether recording a flood's destruction or measuring precipitation over decades, a successful Earth scientist relies heavily on his or her ability to recognize patterns. Patterns in space and time are the keys to many of the great discoveries about how the Earth works. The activities in this module will help you develop your ability to recognize and interpret nature's fundamental patterns by exploring recent scientific data using a geographic information system (GIS). A GIS is a tool for organizing, manipulating, analyzing, and visualizing information about the world using a computer and digital maps.

In Unit 1, you will use data and analysis tools to estimate the amount of water stored in global reservoirs and examine the processes that move water between them. In Unit 2, you will explore precipitation patterns in the U.S. and examine the factors that control them. In Unit 3, you will analyze detailed water use and agricultural data from across the U.S. to discover trends and patterns that affect our daily lives. Finally, in Unit 4, you will investigate the challenges involved in meeting the water needs of a desert city—Tucson, Arizona.

Most of these patterns are presented in maps, which are one of scientists' most important tools. The maps allow you to visually explore the relationships between phenomena such as precipitation and runoff, natural features such as oceans, ice caps, mountain ranges, rivers, and aquifers, and human features such as cities. By investigating a map of precipitation and the areas producing various crops, the patterns in the data will help you understand the devastating effects of droughts and floods, and our dependence on irrigation for a diverse and stable food supply. Maps can even show dramatic evidence of the consequences of over-using the available water supply, including a major metropolitan area that is slowly sinking! Behind each type of feature in the map is a table containing an extensive database of information about that map feature. Careful analysis of these data can reveal patterns in the data that are difficult to discover through visual examination alone.

Planning to learn

Each unit will take you through a well-tested learning process that helps you examine your existing knowledge and build upon it. The first activity will get you thinking about your present knowledge of the major concepts in the unit. It may stimulate questions that you have about the topic. Write these down and, as you learn more, see if you can answer them for yourself.

A GIS Map

This map shows average annual precipitation statistics for the U.S.

Tabular Data

A tabular form of the precipitation data. The yellow highlighted rows show data for the western states.

In the second activity you will explore maps and data looking for patterns. As you examine these patterns, ask questions such as the following:

- Where do they occur? (or *not* occur?)
- Why does this pattern occur here and not elsewhere?
- What might cause this pattern?

The third activity of each unit provides key background information about the major concepts and should help you begin to answer the questions raised above. The readings are brief and explain only the important concepts required to continue or to check your prior answers. In the fourth (and sometimes fifth) activity, you will apply your new knowledge to solve a particular problem. This will help you measure your understanding of the material and prepare you for the final activity where you will be asked to apply the concepts you have learned in the unit to a specific location or situation.

GIS made easier

The purpose of these activities is not simply to learn how to use a GIS, but to use one as a tool to explore and learn about natural processes and features and how they relate to humans and human activities. For this reason, all of the data have been assembled into ready-to-use projects and complex operations have been eliminated or simplified. While it is helpful to have basic computer skills, you do not need experience with ArcView GIS software to complete these activities. Directions for each task are provided in the text, so you will learn to use the tool as you explore with it. It is especially important to pay attention to the tips provided in the margin, such as the *Want to know more?* item at the left. The activities barely scratch the surface of the data that have been provided, and we encourage you to explore the data on your own and make your own discoveries.

Using these materials

Visual cues are used to make the activity directions easier to follow.

- A line preceded by the ▶ symbol is an instruction—something to do on the computer.
- When referring to a tool or button, the name of the tool or button is capitalized and is followed by a picture of that item as it appears on screen—e.g. ...the Identify tool **❶**.
- The ▶ symbol between two boldface words in text indicates a menu choice. Thus, **Theme ▶ Properties** means "pull down the Theme menu and choose Properties."

The most common mistake made when using ArcView GIS software is not activating the correct theme (map layer) before performing an operation. When things do not seem to be going as they should, this is the first thing to check.

The second most common problem is not looking closely enough at the maps to see what is going on. ArcView has several tools and buttons for zooming in and out of the map view that work just like tools you have used in other applications—use them!

Sidebar

Want to know more?

If you would like more information on how to use ArcView GIS, refer to the **Guide to ArcView GIS**, found in the **Docs** folder on the CD.

Theme ▶ Properties *means...*

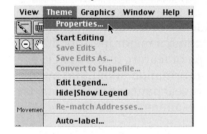

To activate a theme, click on its name in the Table of Contents. Active themes are indicated by a raised border.

To zoom in on an area, click and drag with the Zoom In tool 🔍 to outline the area. **To zoom out**, click anywhere on the map window with the Zoom Out tool 🔍.

Rounding

Some of the numbers you will work with in Exploring Water Resources are quite large. When talking about the amount of water flowing in U.S. rivers and streams, you routinely use values in the billions or even trillions! Most of these numbers are approximations, so it does not make sense to worry about being precise when you are calculating or recording these large numbers. Look at the number written below, and the values of each of the digits. Face it—when you are talking about nearly 149 billion of something, who cares about hundred-thousandths, or even tens of millions?

$$148{,}735{,}992{,}068.95249$$

Throughout Exploring Water Resources, you will be instructed to, for example, *"Round your answer to the nearest 100,000."* Rounding numbers is very simple, if you follow these three steps:

- Look only at the numeral to the right of the place value you are rounding to. For example, when rounding to the nearest thousand, look only at the numeral in the hundreds place.
- If the numeral to the right is 0-4, do not change the number in the rounding place value. If the number to the right is 5-9, add one to the number in the rounding place value.
- Change all of the numerals to the right of the place you are rounding into zeros.

For example, if your number is **319,740,562.85**

To round to the nearest ten million:

1. Find the ten millions digit (1).
2. Look at the number to its right (9). Since it is between 5 and 9, add one to the ten millions digit.
3. Change the numbers to the right of the ten millions digit to zeros. The result is **320,000,000**.

Rounding this number to the nearest…

…million (1,000,000) = 320,000,000 (adding 1 to 319 gives 320)

…hundred thousand (100,000) = 319,700,000

…ten thousand (10,000) = 319,740,000

…thousand (1,000) = 319,741,000

…hundred (100) = 319,740,600

…ten (10) = 319,740,560

…one (1) = 319,740,563

…tenth (0.1) = 319,740,562.9

Rounding and fractions

Rounding fractions works exactly the same way as rounding whole numbers. The only difference is that instead of rounding to tens, hundreds, thousands, and so on, you round to tenths, hundredths, thousandths, and so on. Do not add zeros to the right of the decimal point. In other words, rounding 2.587 to the nearest tenth is 2.6, *not* 2.600.

Getting started

Additional resources

Visit The SAGUARO Project website for updates, references, and links to related websites:

http://saguaro.geo.arizona.edu

The Exploring Water Resources CD contains all the software and data you need to complete the module activities on your own Macintosh® or Windows® computer. If you are using this manual as part of a laboratory course, a computer lab with the necessary software and data files may have already been prepared for you in advance.

What you need to know

The authors of this book assume that you know how to use a computer with either the Macintosh or Windows operating system installed. We will make no attempt to teach these basic skills:

- turning the computer on and, if necessary, logging in as a user;
- navigating the file system to find folders, applications, and files;
- launching applications and opening files; and
- using basic interface elements—opening, closing, moving, and resizing windows, using tools, menus, and dialog boxes, etc.

Installation courtesy

Install the Exploring Water Resources applications and data only on your own personal computer. If someone else owns the machine, you should seek permission before installing anything.

The data files for this module may be accessed directly from the CD or copied to a user-specified drive and directory during the installation process. Thus, in our modules, you will be instructed to launch the ArcView GIS application, then locate and open a specific file. In a lab setting, your instructor will tell you where to find this file.

Minimum system requirements

Macintosh

- ArcView® GIS 3.0a for Macintosh®
- **QuickTime™ 5.0 or newer** and Acrobat® Reader 4.0 or newer (installers included on CD) and Microsoft® Excel for Macintosh®
- 400 MHz or faster PowerPC® CPU running Mac OS 8.0 or newer
- 64 MB total RAM (32 MB of available application RAM)
- CD-ROM drive
- 40 MB available hard disk space for applications (plus up to 225 MB for data, if installed on hard disk)

Mac OS X compatibility

ArcView GIS 3.0a for Macintosh runs in Classic mode under Mac OS X. For best results, restart your computer in OS 9 and install ArcView GIS.

Windows

- ArcView® GIS 3.0–3.3 for Windows®
- **QuickTime™ 5.0 or newer** and Acrobat® Reader 4.0 or newer (installers included on CD) and Microsoft® Excel for Windows®
- 400 MHz or faster Pentium™-class CPU running Windows 98 or newer
- 64 MB total RAM (32 MB of available application RAM)
- CD-ROM drive
- 40 MB available hard disk space for applications (plus up to 225 MB for data, if installed on hard disk)

ArcView 8 compatibility

ArcView project files in Exploring Water Resources cannot be opened directly into ArcView 8.

Macintosh software installation

Macintosh CD-ROM Contents

The **Exploring Water Resources CD** contains the following folders and files.

Readme.txt

License.txt

EWR_Unit_1

    global.apr

    sagemdia.txt

    Data (many files)

    Media (many files)

EWR_Unit_2

    renewable.apr, etc.

EWR_Unit_3

    uswater.apr, etc.

EWR_Unit_4

    ground_water.xls

    tucson_water.apr, etc.

ArcView

    ArcView Installer

Reader

    Acrobat Reader Installer

QuickTime

    QuickTime Installer

Docs

    Guide to ArcView GIS.pdf

    Data Dictionary.pdf

"Can I allocate more memory to ArcView?"

If your computer has more memory available, you can allocate some of the additional memory to ArcView, keeping in mind your other system requirements.

The instructions below are for installing software on your Macintosh computer. If the software has already been installed, such as in a lab setting, skip ahead to the **Using ArcView GIS for Macintosh** section on page ix.

Before you install software

- Insert the **Exploring Water Resources** CD into your CD-ROM drive and read the **Readme.txt** file on the CD for installation updates.
- Disable virus protection software (if installed) and quit any open applications.
- You must have at least 40 MB of free space on your hard drive to install the application. Installing the module data on your hard drive is optional and requires an additional 225 MB of drive space.

Installing ArcView GIS

- Open the **ArcView** folder, double-click the **ArcView Installer** icon, and follow the on-screen instructions.
- When installation is complete, restart your computer and hold down the Command (⌘) and Option keys until the message "Are you sure you want to rebuild the desktop file on the disk..." message appears. Click **OK**. Your computer will rebuild its desktop file and complete the startup process.

Setting ArcView's memory allocation

The default memory allocation for ArcView must be increased to assure trouble-free operation! Follow these steps to increase the application's memory allocation.

- Navigate to where you installed the ArcView GIS application. Look for a folder named **ESRI**—open it and open the folder inside it named **AV_GIS30a**.
- Single-click the **ArcView** application icon 🔍 to select it.
- Choose **File ▶ Get Info**, choose **Memory** from the **Show** pop-up menu.

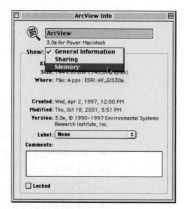

- Enter 32000 for the **Preferred Size** and at least 32000 for the **Minimum Size. (Note—do not include commas!)**

Where is ArcView GIS?

On a Macintosh, ArcView GIS is installed in a folder named **ESRI** on the drive you specified during installation. You will find the ArcView application in the **AV_GIS30a** folder.

Make an alias!

For convenience, you may wish to make an alias of the ArcView application on your desktop. See your documentation for instructions on how to create an alias.

Monitor resolution

The Exploring Water Resources activities were designed to be used on a computer with a monitor resolution of at least 800 by 600 pixels and 256 colors. Be sure to set your monitor accordingly, if necessary.

Changing monitor resolution on a Macintosh

Select **Monitors** from the Control Panels ▸ under the Apple menu. (Your particular computer may say **Monitors and Sound**.) Select the Monitor button and choose the appropriate screen resolution.

Registering ArcView GIS

To register the ArcView GIS application (Macintosh only):

- Double-click the **ArcView** icon 🔍 to launch the application.
- When prompted, enter your name and company (you may leave this blank), then enter the registration number: **708301184404**.
- Quit ArcView.

Installing QuickTime

The investigations in this module use QuickTime for displaying movies and animations. Most Macintosh computers already have QuickTime installed. If yours does not, you should install it.

- Turn on your computer and insert the **Exploring Water Resources** CD into the CD-ROM drive.
- Open the **QuickTime** folder on the CD and double-click the **Quicktime Installer** icon. Follow the on-screen instructions to complete the installation.
- Restart your computer.

Installing Acrobat Reader

Acrobat Reader is used to view and print the files in the **Docs** folder. If Acrobat Reader is already installed on your computer or you do not wish to view or print these documents, you may skip this installation.

- Turn on your computer and insert the **Exploring Water Resources** CD into the CD-ROM drive.
- Open the **Reader** folder on the CD and double-click the **Acrobat Reader Installer** icon. Follow the on-screen instructions to complete the installation.
- Restart your computer.

Using ArcView GIS for Macintosh

Launching ArcView and opening project files

- Double-click the **ArcView** application icon 🔍 to launch ArcView.
- Choose **File ▸ Open**.
- Navigate to the appropriate **EWR_Unit** folder (on your **Exploring Water Resources** CD or installed on your hard drive), select the ArcView project file (the file name ends with .*apr*), and click **Open**.

Closing project files

When you have completed an activity or must stop for some reason,

- Choose **File ▸ Quit**.
- When asked if you want to save your changes, click **No**. (Don't worry if you click **Yes**. The files have been locked to prevent accidentally modifying or erasing them.)

Windows CD-ROM Contents

The Exploring Water Resources CD-ROM contains the following folders and files.

Readme.txt

License.txt

EWR_Unit_1
> *global.apr*
> *sagemdia.txt*
> Data (many files)
> Media (many files)

EWR_Unit_2
> *renewable.apr, etc.*

EWR_Unit_3
> *uswater.apr, etc.*

EWR_Unit_4
> *ground_water.xls*
> *tucson_water.apr, etc.*

ArcView
> *AV32AVC.EXE*

Reader
> *RP505ENU.EXE*

QuickTime
> *QuickTimeInstaller.exe*

Docs
> *Guide to ArcView GIS.pdf*
> *Data Dictionary.pdf*

Monitor resolution

The Exploring Water Resources module was designed to be used with a monitor resolution of at least 800 by 600 pixels and 256 colors.

Changing monitor resolution on a Windows computer

Right-click on the desktop, choose Settings from the popup menu, and click the Settings tab. Set the color palette and desktop area to the appropriate values. For more help, consult your computer's printed or online documentation.

Windows software installation

The instructions below are for installing software on your Windows-based computer. If the software has already been installed, such as in a lab setting, skip ahead to the **Using ArcView GIS for Windows** section on page xi.

Before you install software

- Disable virus protection software (if installed) and quit any open applications.
- Make sure you have sufficient hard disk space for the installation—40 MB for the application. Installing the module data on your hard disk is optional and requires approximately 225 MB of additional hard disk space.
- You may install this version of ArcView alongside older or newer versions by installing it in a different location on your hard disk.

Installing ArcView GIS

- NOTE—This is a 120-day license. Do not install until you need to use it, or your license may expire before the semester is over!
- Insert the **Exploring Water Resources** CD into your CD-ROM drive.
- Open the **ArcView** folder, double-click the **AV32AVC.EXE** icon, and follow the on-screen instructions to complete the installation. *(Note - newer versions of Windows may be configured to hide the three character file extension, so this file would appear as simply AV32AVC.)*
- Restart your computer.

Installing QuickTime for Windows

This module uses QuickTime for displaying animations and movies. To install the QuickTime Player application:

- Turn on your computer and insert the **Exploring Water Resources** CD into the CD-ROM drive.
- Open the **QuickTime** folder on the CD and double-click the **QuickTimeInstaller.exe** icon. Follow the on-screen instructions to complete the installation.
- Restart your computer.

Installing Acrobat Reader for Windows

Acrobat Reader is used to view and print the files in the **Docs** folder. If Acrobat Reader is already installed on your computer or you do not wish to view or print these documents, you may skip this installation.

- Turn on your computer and insert the **Exploring Water Resources** CD into the CD-ROM drive.
- Open the **Reader** folder on the CD and double-click the **RP505ENU.EXE** icon. Follow the on-screen instructions to complete the installation.
- Restart your computer.

Using ArcView GIS for Windows

Launching ArcView GIS for Windows

- On the Windows desktop, click the **Start** button.
- On the **Start** menu, choose **Programs ▶ ESRI ▶ ArcView GIS Virtual Campus Edition ▶ ArcView GIS Virtual Campus Edition**.
- On the next screen, click the **Try** button. This will begin your 120-day license period. Each time you launch the ArcView GIS application, this screen tells you the number of days remaining in the license period.

Where is ArcView?

The installer places ArcView in the **Program Files** folder of the drive you specified, in a folder named **VCampus**.

Create a shortcut

You may want to create a shortcut to ArcView on your desktop, to make it easier to access the program. To find out how to create a shortcut, refer to your Windows documentation.

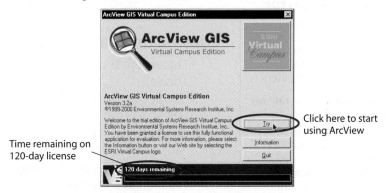

Click here to start using ArcView

Time remaining on 120-day license

Welcome to ArcView GIS

Open project files for **Exploring Water Resources** activities from the **Welcome to ArcView GIS** dialog box.

Opening a project file

- Launch ArcView GIS for Windows (see above).
- In the **Welcome to ArcView GIS** dialog box, click the **Open an existing project** button, then click **OK**.
- Use the **Open** dialog box to locate the file you want to open:
  1) Select the drive where you installed the module files.
  2) Select the directory (folder) containing the project files.
  3) Select the file. ArcView project files end in ".apr."
  4) Click **OK**.

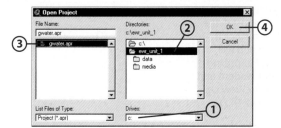

- When the project files have opened, you will see the message shown at left. Answer either "yes" or "no"—it makes no difference.

Closing a project file

When you have completed an activity or must stop for some reason,

- Choose **File ▶ Quit**.
- When asked if you want to save your changes, click **No**. (Don't worry if you click **Yes**. The files have been locked to prevent accidentally modifying or erasing them.)

Accessing the module data

If ArcView and QuickTime 5.0 or newer software are already installed on your computer, you are ready to use the module. You may access the module data directly from the CD-ROM or copy it to your computer's hard drive for increased convenience and performance.

Reading the data files directly from the CD

- Insert the **Exploring Water Resources** CD in your CD-ROM drive.
- Launch ArcView, choose File ▶ Open, navigate to the ArcView project file inside the appropriate **EWR_Unit** folder on the CD.

Copying the data files to your computer

- Insert the **Exploring Water Resources** CD in your CD-ROM drive.
- Copy one or more **EWR_Unit** folders from the CD to your hard disk.
- **Important** - Do not change the name of the **EWR_Unit** folder or any of its contents at any time.
- **Important** - If you are copying an **EWR_Unit** folder into another folder, give the enclosing folder a short name (8 characters or fewer) and do not nest the folder too many levels deep. This will help ensure that ArcView can locate the files correctly.
- **Critical issue for Windows installations only** - There must be *no spaces* in the names of the drive or folders along the path to the **EWR_Unit** folder. If necessary, change spaces in drive and folder names to underscore characters—thus, a **Class Data** folder should be renamed **Class_Data**.
- If the data files get renamed or damaged, delete the entire **EWR_Unit** folder and copy a clean version of the folder from the CD to your hard disk.

Q&A about the 120-day ArcView license on the CD

"My university has an ArcView site license—do I need to use the 120-day version of ArcView on the Exploring Water Resources CD?"

Only if you want to use your own computer to complete the GIS activities—the 120-day version of ArcView is primarily intended for use on students' personal computers.

"When should I install the 120-day version of ArcView?"

Do not install your ArcView GIS software until you need to use it. The ArcView license allows you to install the ArcView application on one computer for 120 days. At the end of that time, the software will no longer function, and it cannot be reinstalled on that computer.

"Can I reinstall the application if it gets damaged or someone accidentally uninstalls it?"

Yes, but it will only function for 120 days from the original installation date.

"What happens at the end of the 120-day license period?"

Your copy of ArcView GIS will stop working. According to the license agreement, you must uninstall the ArcView GIS application, since it will no longer function and will not work if reinstalled. For pricing and information about student and instructor licenses for ArcView, call ESRI telesales at (800) 447-9778.

"What happens if I retake this course or take a different course using another module from the Exploring series?"

You must install your "new" 120-day licensed version of ArcView on a different computer than your original installation.

ArcView troubleshooting

To activate a theme, click on its name in the Table of Contents. Active themes are indicated by a raised border.

"When I _____ in ArcView, nothing (or the wrong thing) happens."

Most ArcView errors are caused by not having the correct theme activated when performing an operation. By activating a theme, you are telling ArcView which data to operate on. If the wrong data are identified, the operation will most likely fail or produce unexpected results.

"ArcView crashes while launching." (Macintosh and Windows)

- ArcView may not have enough memory to work properly (Macintosh only)—see page viii for directions on **Setting ArcView's memory allocation**.
- ArcView may be damaged. You may uninstall and reinstall ArcView at any time within the 120-day license period using the **Exploring Water Resources** CD.

"When I open a project file, ArcView keeps telling me it cannot find files that I know are installed." (Macintosh and Windows)

The project file contains a path to the location of each data file in the project. If you move files or rename any of the files or folders that were installed, ArcView cannot find the files. If you cannot restore the correct file and folder names and locations, delete the **EWR_Unit** folder and re-install it on the hard disk.

"When I open a project file from within ArcView, it tells me it can't find the project file." (Windows only)

There is probably a space in the name of the drive or folder into which you copied the **EWR_Unit** folder. ArcView for Windows cannot locate a file correctly if it encounters a space in the file's pathname.

"ArcView launches and opens the project file, but it crashes later, after several minutes of use ." (Macintosh)

The memory allocation for ArcView is probably too low. See page viii for directions on **Setting ArcView's memory allocation**.

"When I use ArcView's Media Viewer to view movies or animations, I get a message inviting me to upgrade to QuickTime Pro. When I click the "later" button, nothing happens." (Windows only)

This usually occurs only in a lab or other shared environment, when the QuickTime preferences file is not "writable" to all users. The preferences file is called **QuickTime.qtp**; its exact location depends on the version of Windows you are using. Consult your system administrator about making the necessary changes.

"When I try to view a QuickTime movie or animation, I get the message required compressor not found." (Macintosh and Windows)

This message means that you are using an older version of QuickTime that does not support the data compression format used for the movie. Install QuickTime 5 or newer to correct this problem.

"QuickTime movies don't play smoothly." (Macintosh and Windows)

The quality of QuickTime movie playback depends on many factors. Quit any unnecessary applications that are running, and be sure you are using QuickTime 5 or newer. If you are reading files directly from the Exploring Water Resources CD, you should get smoother playback by copying the files to your hard disk.

Updates and resources

For corrections, updates, and additional resources related to **Exploring Water Resources**, visit the SAGUARO Project website at:

http://saguaro.geo.arizona.edu/ewr/

Online help

Two Help menus on a Mac?

On the Macintosh, ArcView adds a second Help menu of its own to the left of the standard Macintosh Help menu. Use this menu to access ArcView's built-in help system.

This module provides all of the directions you need to complete the activities using ArcView GIS. If you wish to explore ArcView on your own, or learn more about ArcView's capabilities, you may wish to consult ArcView's **Help** menu. Choose **Help ▶ How to Get Help...** to learn more about using ArcView's built-in help system.

Printing the Educator's Guide to ArcView GIS

The **Docs** folder contains a handy quick reference manual of the most common tools and techniques used in ArcView, called the Educator's Guide to ArcView GIS. You may view this document on screen or print all or part of it for later reference using the included Acrobat Reader software.

- Turn on your computer and insert the **Exploring Water Resources** CD into the CD-ROM drive.
- Open the **Docs** folder on the CD and double-click the **Guide to ArcView GIS.pdf** icon.
- Use the on-screen controls to scroll through the document, and choose **File ▶ Print** and either print all pages or select a range of pages to print.

The Data Dictionary file

The **Data Dictionary.pdf** file found in the **Docs** folder on the **Exploring Water Resources** CD provides information about the data used in this module, including a description of where the data came from and how they were processed. You do not need this information to complete the activities in the module, but you may find it useful if you are using the data for independent research or just to satisfy your curiosity. Many of the providers listed are good sources for additional data and information.

Unit 1
Global Water Reservoirs

In this unit, you will...

- *explore challenges to providing an adequate amount of fresh, safe water to the world's population;*

- *estimate the volume of water in the oceans, ice caps, and atmosphere;*

- *examine the hydrologic cycle, the time water resides in different global reservoirs, and the challenges of recovering water from each reservoir; and*

- *evaluate the societal and agricultural impacts of global sea level changes caused by melting ice caps.*

NASA/GSFC

Activity 1.1

Global water sources

The water planet

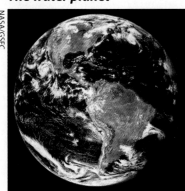

Slightly over three-fourths (75%) of Earth is covered by water in either liquid or solid form.

Water is crucial for all life on Earth. A person can only survive five days without water. Besides the air we breathe, water is the most important resource for sustaining human life. Fortunately, Earth possesses abundant water resources—over three-fourths of its surface is covered by water. This water exists in solid, liquid, and gaseous forms in various storage areas, or *reservoirs*. These reservoirs are found above, on, and below Earth's surface.

The health and economic welfare of the global population depends on a steady supply of fresh, uncontaminated water, so developing ways to extract usable water from different reservoirs is very important. Many existing reservoirs are being depleted by withdrawals and stressed by contamination. Although future management of our water resources will require significant conservation efforts, we may also find ways to make unusable water usable, or to harvest fresh water more economically from reservoirs previously considered impractical. In this activity, you will identify and determine the size and importance of global water reservoirs and think about how your community may utilize these reservoirs.

1. The ocean is the largest reservoir of water. List as many of Earth's other water reservoirs as you can in the table below.

Table 1—Global water reservoir predictions

Global Water Reservoir	Size Rank	Accessibility
oceans	1	

Need a hint?

Don't forget that water may be stored in liquid, solid, or gaseous (vapor) form in these reservoirs!

2. Rank the reservoirs you listed in Table 1 by size. Use 1 to represent the largest reservoir, 2 for the next largest, and so on.

3. Classify the reservoirs you listed based on their accessibility, or how easily fresh water is obtained from them, as either *easy*, *moderate*, or *difficult*.

4. Only fresh, pure water can be used for drinking. Describe three uses of the water that *cannot* be used for drinking.

5. From which reservoirs does your community obtain drinking water?

6. What challenges does your community face in providing a safe and adequate water supply?

Activity 1.2

Measuring global water

Water is essential for life as we know it, and Earth certainly has it in abundance. Water can be found as a solid (ice and snow), a liquid ("water" as we normally think of it), and a gas (water vapor). Oceans, lakes, and rivers are the most obvious places to find water, but it is also in the sky as clouds formed of ice and water droplets. Water is found underground, as a liquid in water-bearing rock formations or *aquifers,* and in frozen soil, or *permafrost.* Even the driest deserts show signs of water, in the form of dried up river channels and lake beds. These occasionally fill with rainwater, sustaining life in what at first seems a barren wasteland. Water is found in many places, yet how much of it is readily available for human use? In this activity, you will look at three global water reservoirs to answer this question.

Earth's oceans: the largest reservoir

The oceans contain the vast majority of Earth's water. In this part of the activity, you will determine the size of the oceans and the amount of water they contain.

▶ Launch ArcView and locate and open the **gwater.apr** file.

▶ Open the **Earth's Oceans** view. It should look like this:

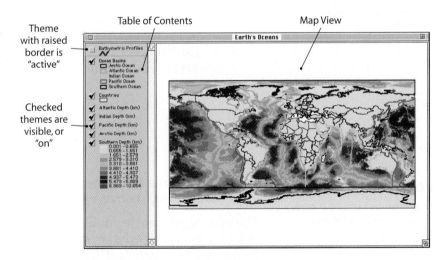

Four oceans or five?

The division of Earth's oceans into distinct basins is somewhat arbitrary. In the past, only four oceans were named. Recently, however, a fifth was described—the Southern Ocean. Many sources still list only the original four, the Atlantic, Indian, Arctic, and Pacific Oceans.

Bathymetry

Bathymetry is a measure of the water depth in basins such as lakes, oceans, and rivers. The root, **bathy**, comes from a Greek word meaning *depth*. **Metry** comes from another Greek word meaning *to measure*.

Scientists divide Earth's oceans into five basins, which are outlined in different colors in the **Ocean Basins** theme. Ocean depth, or *bathymetry*, is represented by five themes: **Atlantic Depth (km)**, **Pacific Depth (km)**, **Indian Depth (km)**, **Arctic Depth (km)**, and **Southern Depth (km)**. Each uses the same legend, shown beneath the **Southern Depth (km)** theme in the Table of Contents. Ocean depth is indicated by varying shades of color.

Rounding

If you need a refresher on rounding, see page *v*.

Bathymetric profiles are cross-sectional views of ocean basins. Click the Media Viewer button 🖼 and choose **Profile Movie** to watch a movie that explains bathymetric profiles.

To turn a theme on or off, click its checkbox in the Table of Contents.

To activate a theme, click on its name in the Table of Contents. Active themes are indicated by a raised border.

To use the Hot Link tool, position the *tip* of the lightning bolt cursor over the feature and click.

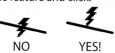

NO   YES!

Plate tectonics

New lithosphere is created at ocean ridges and older lithosphere is recycled at trenches.

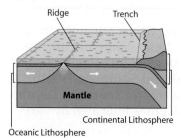

Oceanic Lithosphere

These themes are quite large, so they display slowly. In general, display only the bathymetry themes you need at a given time. For now, however, leave all five turned on.

Study the legend for the bathymetry themes.

1. How deep are the deepest portions of the oceans? Round to the nearest 0.1, and don't forget to include units.

Next, you will examine bathymetric profiles to get a sense of the overall shape of the ocean basins.

▶ Turn on and activate the **Bathymetric Profiles** theme.

This theme shows twelve line segments drawn across different regions of the world. Each segment is hot linked to a profile that shows the ocean depth along that line segment.

▶ Use the Hot Link tool ⚡ to click on a line segment.

An image window will open, showing the profile of the ocean bottom along that line segment. If an image window does not open, make sure the **Bathymetric Profiles** theme is active.

Use the **Bathymetric Profiles** and ocean depth themes to answer the following questions.

2. Is the depth uniform in all of the oceans? Explain.

3. In general, where are the ocean basins deepest? Describe your observations from the bathymetric profiles.

▶ Turn off the **Bathymetric Profiles, Ocean Basins** and all ocean bathymetry themes except for the **Atlantic Depth (km)** theme.

The process of *plate tectonics* is responsible for the patterns in ocean bathymetry. The deepest parts of the oceans are *trenches*, which are formed as the ocean floor (or *lithosphere*) pushes deep into the Earth and returns to the mantle. At the high *ridges* near the middle of the ocean basins, injections of hot, molten rock along the ridge create new lithosphere. As this new lithosphere is created at ridges, old lithosphere is recycled at trenches.

Next you will determine the maximum and average depth of each of the five oceans. You will begin by examining the Atlantic Ocean.

▶ Activate the **Atlantic Depth (km)** theme.

▶ Click the Open Theme Table button 🖩 to open the **Atlantic Depth (km)** theme table.

Each ocean basin is divided into many small areas. The theme table gives the depth, in kilometers, of each of these areas. You will calculate the average (or mean) depth of each ocean.

The ▶ symbol between two boldface words in text indicates a menu choice. Thus, **Field ▶ Statistics...** means *pull down the **Field** menu and choose **Statistics...** from the menu.*

- ▶ Click the **Depth (km)** field heading. The field name will be highlighted gray to show that it is selected.
- ▶ Choose **Field ▶ Statistics...**

After the statistics are calculated (be patient—it may take a while), a window named **Statistics for Depth (km) field** will appear.

- ▶ In the statistics window, read the maximum depth (**Maximum**) and the mean depth (**Mean**).

4. Round the maximum and mean depths for the Atlantic Ocean to the nearest 0.1 km and record them in Table 1.

Table 1—Ocean volume

Ocean Name	Maximum Depth (km) (statistics window)	Mean Depth (km) (statistics window)	Surface Area (km²) (Identify tool)	Volume (km³) (= Mean Depth x Surface Area)
Atlantic				
Indian				
Pacific				
Arctic				
Southern				
Total				

- ▶ Close the statistics window and the **Atlantic Depth (km)** theme table and turn off the **Atlantic Depth (km)** theme.
- ▶ Repeat this process to record the maximum and mean depths for each of the four remaining oceans in Table 1. *Remember to turn on and activate the correct ocean depth theme before opening the theme table and calculating statistics, and to round the depths to the nearest 0.1 km.*
- ▶ When you are finished, close all tables and turn off all ocean depth themes.

Next you will estimate the volume of each ocean by multiplying the ocean's surface area by its mean depth.

- ▶ Turn on and activate the **Ocean Basins** theme.
- ▶ Using the Identify tool ❶, click the Atlantic Ocean in the map view to display information about the Atlantic Ocean.
- ▶ Read the surface area of the ocean, labeled **Area (km2)**, in the **Identify Results** window.

5. Round the surface area of the Atlantic Ocean to the nearest 100,000 km² and record it in Table 1.

- ▶ Close the **Identify Results** window.

Calculating volume

You can calculate the volume of a liquid by multiplying its surface area by its depth.

Volume = Area × Depth

Where's the view?

If you can't find the list of views in the project window, click the Views icon in the project window's table of contents.

Click here.

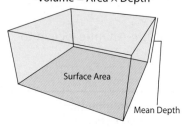

Location map

▶ Repeat this process to record the surface area for each of the four remaining oceans in Table 1. *Remember to round each area to the nearest 100,000 km².*

6. Add together the surface areas of the five oceans and record the total area of the oceans in Table 1.

▶ Calculate the volume of each ocean by multiplying the ocean's mean depth by its surface area.

7. Record the volume you calculated for each ocean in Table 1. Round to the nearest 1,000,000 km³.

8. Add together the volumes of the five oceans and record the total ocean volume in Table 1.

▶ Close the tables and the **Earth's Oceans** view.

Frozen water in continental ice caps

Another important water reservoir is the ice caps. Land masses near both the north and south poles are cold enough throughout the year that what little precipitation occurs there falls as snow. Over millions of years, the snow in each location has compacted into a solid, permanent ice cap a thousand or more meters thick. A significant volume of Earth's fresh water is temporarily locked up in these ice caps.

▶ Open the **Continental Ice Caps** view (see *Where's the view?*, at left).

This view shows the countries of the world as well as two themes that outline the two continental ice caps: **Antarctica Ice Thickness (km)** and **Greenland Ice Thickness (km)**.

Study the legend for the two themes. Each uses the same legend, which is shown beneath the **Antarctica Ice Thickness (km)** theme.

9. What color represents the thickest ice? What thickness range does it represent? Include units and round to the nearest 0.1 km.

As with the oceans, you will first determine the mean thickness of each of the two ice caps. Then, you will determine their surface areas and estimate their volumes.

To determine the mean thickness of the Antarctica ice cap:

▶ Activate the **Antarctica Ice Thickness (km)** theme.

▶ Click the Open Theme Table button 🔳 to open the **Antarctica Ice Thickness (km)** theme table.

▶ Click the **Thickness (km)** field heading. The field name will be highlighted gray to show that it is selected.

▶ Choose **Field ▶ Statistics...** to calculate statistics for the thickness field.

After a few moments, a window titled **Statistics for Thickness (km) field** will appear.

▶ In the statistics window, read the mean thickness of the Antarctica ice cap (**Mean**).

10. Record the mean thickness in the **Mean Thickness (km)** column in Table 2. Round to the nearest 0.1 km.

Table 2—Ice cap volume

Ice cap	Mean Thickness (km) (From statistics window, to nearest 0.1 km)	Area (km²) (From status bar, to nearest 100,000 km²)	Ice Volume (km³) (= Mean Thickness x Area, to nearest 100,000 km³)	Water Volume (km³) (= Ice Volume x 0.9, to nearest 100,000 km²)
Antarctica				
Greenland				
Total				

▶ Close the statistics and theme table windows.

▶ Repeat these procedures to find the mean thickness of the Greenland ice cap and record your results in Table 2.

Next you will determine the ice caps' areas by tracing an outline around each ice cap. The ice caps are at or near Earth's poles, so you will switch the map to a different projection to make them easier to trace.

▶ Click the Orthographic Projection button ⊙.

Now you are viewing Earth as a globe. You should be able to see Antarctica at the bottom and Greenland near the top.

▶ Use the Rotate Projection buttons ← ↑ ↓ → to rotate the view of the Earth until the Antarctica ice cap is centered in the view. Each click rotates the globe 30°. To rotate from the equator to one of the poles, click the appropriate arrow button three times.

▶ Use the Zoom In tool ⊕ to zoom in on the Antarctica ice cap.

Where's the Draw Polygon tool?

The Draw Polygon tool ⬠ is located in the pop-up group at the end of the row of tools.

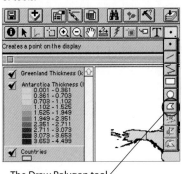

The Draw Polygon tool

▶ Use the Draw Polygon tool ⬠ (see sidebar) to trace a rough outline of the ice cap. Click on a starting point and click again each time you need to change direction. When you get back to the starting point, double-click to finish the polygon. If you don't like the way your polygon looks, delete it and start over.

Read the area of the Antarctica ice cap from the status bar, found either above or below the view window. The units are square kilometers (km²).

Read area here. (Your result will be different.)

Segment Length: 179.07 km Perimeter: 8,149.61 km Area: 15,069,087.28 sq km

11. Record the area in the **Area (km²)** column in Table 2. Round to the nearest 100,000 km².

▶ Choose **View ▶ Full Extent** to view the entire globe.

▶ Rotate the globe until Greenland is centered in the view, then repeat the above procedure to measure and record the area of the Greenland ice cap in Table 2.

▶ For each ice cap, multiply its mean thickness by its area to determine the volume.

12. Record the volume in the **Ice Volume (km³)** column in Table 2. Round to the nearest 100,000 km³.

Polar deserts

Although covered in ice, Antarctica is considered a desert because its mean annual precipitation is only about 5 centimeters (2 inches)—about the same as the Sahara Desert. The main difference is that water in the Sahara evaporates rapidly, while ice in Antarctica evaporates very slowly.

Other ice on Earth

In addition to the continental ice caps that cover Greenland and Antarctica, large areas of Earth are covered by glaciers and sea ice.

As water freezes, it expands. If the ice caps melted, the resulting liquid water would occupy only 90% of the volume of its solid form.

▶ Calculate the volume of liquid water in the ice caps by multiplying the volume of each ice cap by 90% (0.9).

13. Record the volume of liquid water in each ice cap in the **Water Volume (km³)** column in Table 2. Round to the nearest 100,000 km³.

▶ Calculate the total volume of ice and the total equivalent volume of water for the continental ice caps.

14. Record the total volumes in the last row in Table 2.

15. Speculate on why the fresh water in the ice caps is generally considered unavailable for human use.

▶ Close all tables and views.

Water vapor in the atmosphere

The third global water reservoir you will examine is the atmosphere. Clouds in the atmosphere are made up of ice crystals and water droplets, but the air itself also contains water in the form of invisible water vapor.

▶ Open the **Global Water Vapor** view.

Where's the view?

Remember, if you can't find a list of views in the project window, click the Views icon in the project window's table of contents.

This view shows the mean, or *average* amount of water vapor in the atmosphere during a year. The **Mean Water Vapor** theme is color-coded according to the amount of water vapor in a column of air reaching from the surface of the Earth to the top of the atmosphere, a height of about 50 km. The units are cubic meters of water per square kilometer of surface area. Examine this theme's legend briefly, and then study the view.

Columns of air

You can picture a column of air as a tall, transparent building reaching from Earth's surface to the top of the atmosphere.

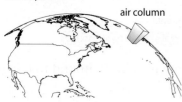

air column

Since Earth is a sphere, the area of the top of the air column is slightly larger than its area on the surface.

16. Briefly describe the distribution of water vapor in the atmosphere. Which regions have large quantities of water vapor and which have smaller quantities?

▶ Turn on the **Mean Temperature (C)** theme.

The **Mean Temperature (C)** theme shows the mean temperatures of the Earth's surface.

17. Where is the mean temperature highest? How does the distribution of water vapor compare to temperature?

To calculate the volume of water in the oceans and the ice caps, you multiplied their surface area by their average depth or thickness. For water vapor, you will use a slightly different technique. The volume of water in each column of air in the **Mean Water Vapor** theme has been calculated for you. To determine the total volume, you will simply add the volumes of the individual air columns together.

▶ Activate the **Mean Water Vapor** theme.

▶ Click the Open Theme Table button 🔲 to open the **Mean Water Vapor** theme table.

Each row of the table represents a small region in the view. The first field (column) of the table is the density of water vapor in the air column above that region, in m^3/km^2. The second field is the area of the region in km^2. The third field is the product of the first two fields, or the total amount of water vapor in the atmosphere over each region, in km^3.

To sum the volumes across the entire Earth, follow these steps:

▶ Click the **Water Vapor Volume (km3)** field heading. The field name will be highlighted gray to show that it is selected.

▶ Choose **Field ▶ Statistics...** to calculate statistics for the water vapor volume field.

In the **Statistics for Water Vapor Volume (km3) field** window that appears, the total amount of water vapor in the atmosphere is given as the **Sum**.

18. How much water vapor is stored in the Earth's atmosphere? Include proper units and round to the nearest 1000 km^3.

19. How does water leave this reservoir?

20. Speculate on the challenges we face in extracting usable amounts of fresh water directly from the atmosphere.

Total water budget

Earth's water is stored in places other than the oceans, continental ice caps, and the atmosphere. To get an overall picture of the global water budget, these other reservoirs must be considered. Table 3 below lists Earth's water reservoirs. The volumes of the reservoirs you have not examined are provided for you in the table.

21. Fill in the **Water volume (km³)** column for the oceans (from Table 1), continental ice caps (from Table 2), and the atmosphere (from question 18) in Table 3 below.

Table 3—Total volume of water on Earth

Reservoir	Water volume (km³)	Percent of total water
Ground water	8,340,000	0.62%
Freshwater lakes	125,000	0.0093%
Inland seas	104,000	0.0078%
Soil moisture	67,000	0.005%
Rivers	1250	0.000093%
Oceans		
Continental ice caps		
Atmosphere		
Total		100%*

* Table does not include the water contained in Earth's biomass, so the sum of the source percentages will not add to exactly 100%.

22. Calculate the total volume of water on Earth by adding the individual volumes of each of the reservoirs and record this number in the last row in the table. Round to the nearest 1,000,000 km³.

23. Using the total amount of water, calculate the percentage each reservoir contributes and record it in the **Percent of total water** column in the table. Round to the first two non-zero digits.

24. Based on the table, what percentage of the world's potable water (fresh water that is fit to drink) is directly accessible in surface reservoirs?

25. Compare the information in Table 3 above to the predictions you made about global water reservoirs in Activity 1.1. Which reservoirs did you over-estimate and which did you under-estimate?

Calculating a percentage

Divide the amount of water a reservoir contains by the total amount of water on Earth. Multiply by 100.

For example, if the ice caps contain 25,000,000 km³ of water and Earth has 1,300,000,000 km³ of water total, the percentage of water in the ice caps would be

$$\% = 25{,}000{,}000 / 1{,}300{,}000{,}000 \times 100$$
$$= 0.019 \times 100$$
$$= 1.9\%$$

Activity 1.3
Utilizing global water reservoirs

The water planet

Imagine that you could travel far out into the Solar System, and look back at Earth. What would you see? As this photo of Earth from space shows, you would see little more than a pale blue dot. The size of the dot clearly illustrates how small we are in the overall scheme of things, but it is the

Figure 1. Earth from a distance of 6.5 billion kilometers, as seen by the Voyager 1 spacecraft on February 14, 1990.

color, visible even at this incredible distance, that reveals the unmistakable presence of water. The abundance of free-flowing liquid water makes Earth unique in our solar system.

The oceans hold an incredibly vast and deep expanse of water, and there is also water in the air we breathe and in the ground beneath our feet. Despite the abundance of water on this planet, over 1 billion people do not have access to clean, safe water for consumption, sani-

tation and hygiene. In developing countries, 90% of wastewater is returned untreated into rivers and streams that supply water for drinking and hygiene. As a result, over 2 million people die each year from water-borne diseases including cholera, typhoid, dysentery, infectious hepatitis, and giardiasis. Although the statistics are most alarming for developing countries, all countries face the problem of supplying sufficient amounts of clean, safe water for drinking and other uses. Next, you will look at how water cycles through Earth's reservoirs, and examine how we use the reservoirs to meet our water needs.

The hydrologic cycle and residence times

Water exists in the atmosphere, oceans, ice caps, and on land. It moves between these reservoirs in a process called the hydrologic cycle, as shown in Figure 2. The movement of water may take many paths between reservoirs and frequently changes its physical state through evaporation (liquid to gas), condensation (gas to liquid), freezing (liquid to solid), melting (solid to liquid), and sublimation (gas to solid, or the reverse). The movement of water in the hydrologic cycle is driven by solar energy, assisted by gravity. Heat energy from

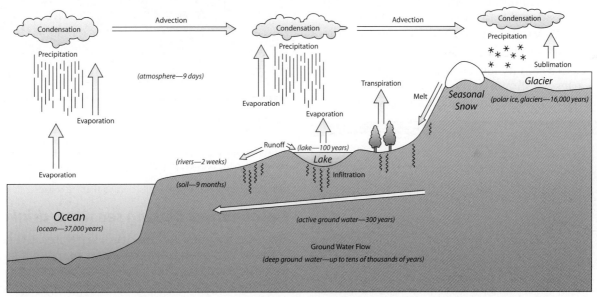

Figure 2. The hydrologic cycle in action. Numbers in parentheses represent typical residence times for water molecules.

the sun causes water to evaporate from oceans, rivers, soils, and even plants and move into the atmosphere. Warm air currents cause water vapor to rise, cool, and condense. The force of gravity pulls all objects toward the center of the Earth; as water condenses out of the atmosphere in the form of precipitation, gravity pulls it toward the ground. In addition to the vertical movement of water described above, the uneven heating of Earth's surface by the sun generates horizontal air movement in the atmosphere, a process known as **advection.** When water reaches Earth's surface, gravity dictates the movement of runoff as it travels across the surface into lakes, rivers, and streams. Gravity also determines the movement of water through aquifers beneath the surface.

Figure 2 illustrates the processes that move water between the major global reservoirs in the hydrologic cycle. It also identifies the **residence time** for each reservoir, the average time a water molecule spends in each reservoir. In the atmosphere, the residence time is just nine days. This means that an average water molecule in the atmosphere precipitates out and is replaced by a "new" one every nine days. Residence times in the oceans, ice caps, and ground water are much longer, on the order of thousands to tens of thousands of years.

The ocean as a water source

Oceans contain about 97% of Earth's water. Unfortunately, we cannot drink the water from this reservoir or use it to grow food because it is too salty. Precipitation dissolves minerals from the rocks on land and transports them to the oceans. When water evaporates from the oceans, the dissolved minerals are left behind.

The ocean's salinity level (the amount of dissolved salt) varies, as shown in the global salinity map in Figure 3. The salinity is generally lower in the

polar regions where glacial meltwater and low evaporation rates combine to dilute the surrounding ocean waters. Similarly, lower salinity is found along coastlines where rivers empty fresh water into the oceans. In contrast, elevated salinity levels occur where evaporation rates are high and precipitation rates are low, as in the Mediterranean Sea, the Red Sea, and the Gulf of Mexico.

Normal ocean salinity is approximately 35 parts per thousand, which means that a thousand pounds of sea water contain about thirty-five pounds of salt. The world's oceans therefore contain approximately 50 million billion tons of salt, enough to cover the land surface of Earth to a depth of around 500 feet.

Why can't we drink sea water?

The most common salt in sea water is sodium chloride (table salt). Sodium is an important nutrient, and is involved in a variety of bodily processes. Normally, the sodium level in a person's bloodstream is in balance with the sodium levels within the individual cells in the body. However, when a person consumes salt water, the sodium levels in the bloodstream increase. This difference causes water to move from the cells into the bloodstream to restore this balance by diluting the excess sodium. The movement of water out of the cells causes them to shrink and stop functioning, a potentially deadly situation. Figure 4 illustrates the effect of sodium levels on red blood cells.

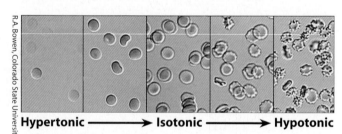

Figure 4. The effect of sodium levels on red blood cells. In **isotonic** blood, the sodium levels of the cells and the surrounding liquid, or plasma, are at equilibrium, and cells appear normal. In **hypertonic** blood, the concentration of sodium in the cells is higher than in the plasma. Water diffuses into the cells, causing them to expand and burst. In **hypotonic** blood, the concentration of sodium in the plasma is higher than in the cells. Water diffuses out of the cells, causing them to collapse.

Desalination: Making sea water drinkable

Because humans cannot drink saltwater, the salt must be removed from the water to make it suitable for drinking. This process occurs naturally over the oceans, where solar energy evaporates water from the surface of the oceans. As the water

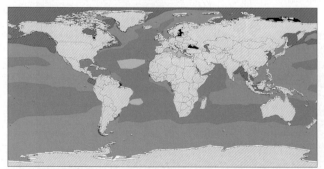

Figure 3. Global salinity map. Low salinity is represented by blue, and higher salinity by green and yellow. For a complete explanation of why the oceans are salty, see http://www.palomar.edu/oceanography/salty_ocean.htm.

evaporates, it leaves the minerals behind. This explains why all precipitation is fresh water, and why the oceans are salty. This process of removing salt from ocean water is called **desalination**. Today there are more than 12,000 desalination plants in operation worldwide. Most are located in the Middle East, producing about 5 billion gallons of water per day. Saudi Arabia alone produces 800 million gallons of fresh water per day from 27 large desalination plants.

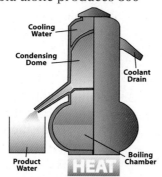

Figure 5. Diagram of basic distillation apparatus. Adapted from http://www.desware.net

Two basic techniques are used to remove salt from ocean water: **reverse osmosis**, in which the water is pumped through a membrane to filter out salt, and **evaporation** or **distillation**. Figure 5 illustrates the basic distillation process.

Desalination is expensive, due to the amount of power required to pump and filter the water. In California, water produced by desalination plants costs about $10 per thousand gallons, which seems pretty inexpensive. However, compared to the cost of Colorado River water at about $0.01 per thousand gallons, it is not surprising that most of these plants sit unused, except in emergencies.

Technological improvements may lower the cost of desalinating water. A new desalination plant scheduled to begin operation in Tampa Bay, Florida is designed to produce water for about $2 per thousand gallons. Still, we are a long way from turning the world's oceans into an economically viable source of fresh water.

Ice caps as a water source

As the largest reservoir of fresh water on Earth, the continental ice caps have long been considered a potential source of drinking water. Every year, about 40,000 icebergs break off the Greenland ice cap. Some of these are pushed southward by the Labrador Current into the north Atlantic Ocean. Since the average Arctic iceberg is about the size of a fifteen-story building and contains the equivalent of about 30 million gallons of water, icebergs seem like a simple solution to water shortages. Antarctic icebergs, like the one shown in Figure 6, are even larger.

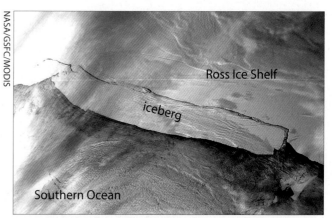

Figure 6. A massive iceberg known as B-15 broke off Antarctica's Ross Ice Shelf in March 2000. Among the largest ever observed, the iceberg is about 11,000 square kilometers—nearly the size of Connecticut.

Schemes to lasso and tow icebergs have been discussed for years, but have generally been dismissed due to economic and environmental considerations. Recently, an Australian scientist has proposed wrapping Antarctic icebergs in plastic to control melting and using natural currents to assist in the transportation process, so the idea continues to have supporters.

Figure 7. Icebergs are routinely towed to avoid collisions with oil drilling platforms and ships. Although currently impractical, recent research shows that icebergs could someday provide fresh water for dry coastal areas.

Atmospheric vapor as a water source

The most direct way to collect atmospheric water vapor is to harvest precipitation before it reaches the ground and is lost to infiltration or runoff. A common approach is to catch precipitation flowing from rooftops and store it in ponds or tanks for later use. This is an effective way to supplement existing water supplies, but does not provide reliable quantities and cannot be used for drinking without expensive filtering systems.

In areas with little or no precipitation, it may be possible to extract usable quantities of water vapor directly from the atmosphere. This is often done by mimicking processes used by other organisms to extract atmospheric water vapor. For example, tropical rainforest plants called **epiphytes** that live attached to the branches of trees far above the forest floor use specialized scales on their leaves to capture and absorb water from the humid air.

S Mukerji, IDRC

Figure 8. Resembling huge volleyball nets, these fog collectors provide fresh water for hundreds of residents of the coastal fishing village of Chungungo, Chile.

Humans can also capture atmospheric water using simple devices called fog water collection systems. These devices are basically large, fine nets that catch and collect the small airborne droplets of water that make up fog. Figure 8 shows one of these systems currently in use in Chile's Atacama Desert that produces about 2500 gallons of water per day. This system doubled the amount of water available to the community and eliminated the need for hauling water in by truck.

A scientist in Germany has recently developed a system that uses new materials that "grab" airborne water, greatly increasing the amount of water vapor caught for a given surface area. Fog water collection represents a promising solution for obtaining water for small, isolated communities.

Global warming

Global warming—in which the Earth's mean temperature rises due to many different factors over geological time—is a large concern today. The phenomenon is a complex issue, both scientifically and politically, and there have been numerous studies and counter-studies that support and refute the claim that it is occurring. One thing we know for certain is that one consequence of global warming is increased melting of the polar ice caps. In the next activity you will investigate the impact of such an event.

Questions

1. The total global population is about 6.2 billion. What percentage of the world's population does not have access to a reliable source of fresh, clean water?

2. In which global reservoir does water reside for the longest time? In which reservoir does it reside the shortest time?

3. What are two processes for desalinating water?

4. Is it practical to use the ice caps or sea ice as sources of fresh water? Explain.

5. How is water harvested from the atmosphere?

Activity 1.4

What if the ice caps melted?

Introduction

Global warming—a situation in which the Earth's mean temperature rises over long periods of time—is of great concern today. In this activity, you will examine the potential consequences of an extreme global warming event in which the Antarctica and Greenland ice caps melt.

1. What would happen if the ice caps melted? Speculate on how it would affect our land, economy, and society.

Sea level and melting ice caps

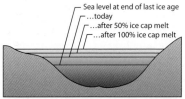

Sea level at end of last ice age
...today
...after 50% ice cap melt
...after 100% ice cap melt

Changes in sea level caused by changes in global ice caps (not to scale).

The change in mean sea level

To estimate the change in sea level that would occur if the continental ice caps melted, you will determine the volume of the ice caps and then calculate how that water would change the depth of the oceans.

2. In **Activity 1.2**, you determined the total volume of water in the ice caps (Table 2, page 9) and the total surface area of the oceans (Table 1, page 7). Write those values below. Include the units.

   a. Volume of water in ice caps =

   b. Surface area of oceans =

If the ice caps were to melt, their water would spread out across the surface of the oceans, raising sea level. The rise in sea level would actually increase the surface area of the oceans as water spread out on to what is presently dry land. To simplify this calculation, however, assume the change in the oceans' surface area is small enough to be ignored.

Calculating sea level rise

Change in sea level =

$$\frac{\text{Volume of ice caps}}{\text{Surface area of oceans}}$$

How big is a meter?

A meter is about 39 inches—slightly more than 1 yard.

▶ Calculate the rise in sea level by dividing the volume of water in the ice caps by the total surface area of the ocean. To convert from kilometers to meters, multiply the result by 1000.

3. By how much would sea level rise if the ice caps were to melt?

Effects of the rising sea level

Now that you have a rough estimate of the change in sea level due to the complete melting of the ice caps, you can examine the effects of a sea level rise on global land masses.

▶ Launch ArcView and locate and open the **gwater.apr** file.
▶ Open the **Melting Ice Caps** view.

This view shows the countries of the world and the boundaries of the ocean projected after a 10-meter, 30-meter, and 60-meter rise in sea level. The boundary for the 60-meter rise includes the 30-meter rise and the 10-meter rise.

4. On the map below, indicate the two regions that would experience the greatest flooding from a 60-meter rise in sea level.

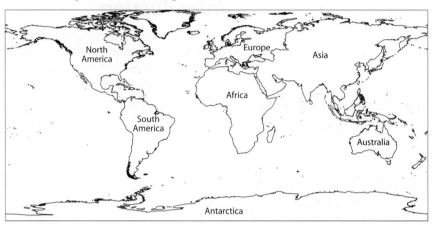

To turn a theme on or off, click its checkbox in the Table of Contents.

To activate a theme, click on its name in the Table of Contents. Active themes are indicated by a raised border.

Which state is that?

If you need help identifying states, make sure the **US States** theme is active and on. Using the Identify tool, click on a state and read its name in the Identify Results window.

▶ Use the Zoom In tool to zoom in on the contiguous 48 United States (excluding Alaska and Hawaii).

5. Near which U.S. coast (east or west) would you rather be living if the ice caps were to melt? Why?

▶ Turn on and activate the **US States** theme.

6. After a 60-meter sea level rise, which states are completely underwater?

▶ Choose **View ▶ Full Extent** to view the whole world.
▶ Turn off the **10-meter Rise**, **30-meter Rise**, and **60-meter Rise** themes and turn on the **Affected Population** theme.

The **Affected Population** theme shows the global population living within the region affected by a 60-meter increase in sea level. The theme consists of a number of rectangles, each of which is color-coded according to its population. Areas with fewer than 10,000 inhabitants are not shown.

▶ Use the Zoom In tool to explore the population data.

7. On which continents will the most people be affected by a 60-meter rise in sea level? How does this compare to your answer for question 4?

▶ Choose **View ▶ Full Extent** to view the entire map.

Next, you will estimate the total population affected, the total number of cities inundated by floodwaters, and the amount of cropland flooded if sea level were to rise 10, 30, or 60 meters.

Total flooded area

Determine the total area flooded for each rise in sea level, starting with the **10-meter Rise** theme.

▶ Turn on and activate the **10-meter Rise** theme.
▶ Click the Open Theme Table button 🖽 to open the **10-meter Rise** theme table.
▶ Click the **Area (km2)** field heading. The field name will be highlighted gray to show that it is selected.
▶ Choose **Field ▶ Statistics....**

In the statistics window, the total area flooded is given as the **Sum**.

8. Record the total area in the **Total flooded area (km²)** column of Table 1. Round to the nearest 100,000 km².

Rounding

If you need a refresher on how to round, see the examples on page v.

Table 1—Impacts of sea level rises of 10, 30, and 60 meters

Sea level rise	Total flooded area (km²) to nearest 100,000	Total population affected to nearest 100,000,000	Number of cities flooded	Total cropland flooded (km²) to nearest 100,000
10 meters				
30 meters				
60 meters				

▶ Close the **10-meter Rise** theme table.
▶ Repeat the above steps to find the total area flooded by a 30-meter and 60-meter rise in sea level and record your results in Table 1. Be sure to activate the appropriate theme each time.
▶ When you are finished, close all theme table windows.

Affected population

Next, estimate the affected population in each of the flooded areas.

▶ Activate the **Affected Population** theme.
▶ Choose **Theme ▶ Select By Theme....**
▶ Choose from the pop-up menus to make the dialog box read "Select features of active themes that **Intersect** the selected features of **10-meter Rise** (do *not* select "Cropland, 10-m Rise"), and click the **New Set** button.

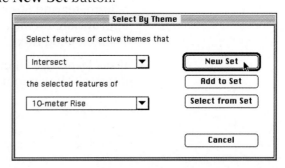

The densely populated areas that would be flooded by a 10-meter rise in sea level will be highlighted in the map.

▶ Click the Open Theme Table button ▦ to open the **Affected Population** theme table.

▶ Click the **Population** field heading. The field name will be highlighted gray to show that it is selected.

▶ Choose **Field ▶ Statistics....**

The total number of people affected by a 10-meter rise in sea level is reported in the statistics window as the **Sum**.

9. Round the total population affected to the nearest 100,000,000 and record it in Table 1.

▶ Repeat these steps to determine the total population that would be affected by a 30-meter and 60-meter rise in sea level and record these totals in Table 1. Be sure to choose the appropriate theme in the lower pop-up menu of the Select By Theme dialog box.

▶ Turn off the **Affected Population** theme and close all theme table windows.

Flooded cities

Next, you will determine the number of cities affected by flooding.

▶ Turn on and activate the **Major Cities** theme.

The **Major Cities** theme includes cities with populations greater than 20,000, capitals, and small but geographically-important towns. The size of the circle represents the size of the city's population. Determining the total number of cities affected is similar to estimating the affected populations.

▶ Choose **Theme ▶ Select By Theme...**

▶ In the dialog box, set the pop-up menus so that the sentence reads "Select features of active themes that **Intersect** the selected features of **10-meter Rise**."

▶ Click the **New Set** button.

▶ Click the Open Theme Table button ▦ to open the **Major Cities** theme table.

▶ Read the number of flooded cities in the status bar.

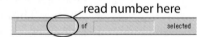

10. Record the number of cities in Table 1.

▶ Repeat these steps to determine the number of cities flooded by a 30-meter and 60-meter rise in sea level. Be sure to choose the appropriate theme in the lower pop-up menu of the Select By Theme dialog box.

11. If sea level rose 10 meters, would a major city in your state be flooded? How might this affect you?

▸ Turn off the **Major Cities** theme.

Flooded croplands

Finally, you will determine the amount of cropland affected by flooding, using a series of themes that show the intersections between the flooding zones and the world's croplands.

▸ Turn on and activate the **Cropland, 10-m Rise** theme.

▸ Click the Open Theme Table button 🖿 to open the **Cropland, 10-m Rise** theme table.

▸ Click the **Area (km2)** field heading. The field name will be highlighted gray to show that it is selected.

▸ Choose **Field ▸ Statistics....**

The total area of cropland flooded by a 10-meter rise in sea level is given as the **Sum**.

12. Record the total area of cropland flooded in Table 1. Round to the nearest 100,000 km².

▸ Close the statistics window.

▸ Repeat the above steps to find the total area of cropland flooded by a 30-meter and 60-meter rise in sea level. Be sure to turn on and activate the appropriate cropland theme each time.

By themselves, these figures on area, population, and cropland affected by a rise in sea level are somewhat hard to assess. It is useful to compare the figures to the totals for each category and calculate the percentage of land, population, and cropland affected. These totals are given in Table 2 below.

Table 2—Global statistics

Total land area	147,000,000 km²
Total population	6,200,000,000 people
Total major cities	606 cities
Total cropland area	23,000,000 km²

Calculating percentages

To calculate percentages for Table 3, divide the value from the corresponding cell in Table 1 by the appropriate total in Table 2, and multiply by 100. For example, for a 10-meter rise in sea level:

% Land area affected =

$$\frac{5,400,000 \text{ km}^2}{147,000,000 \text{ km}^2} \times 100 = 3.67\%$$

Round to nearest 1% = 4%.

13. Use the totals provided in Table 2 to calculate the percentage affected for each category for a 10-, 30-, and 60-meter rise in sea level and complete Table 3. Round all results to the nearest 1%.

Table 3—Effects of rising sea level

Sea level rise	% Area affected	% Population affected	% Major cities affected	% Cropland affected
10 meters				
30 meters				
60 meters				

Plotting your results on graphs will help you interpret them. First, you will examine the relationship between the percentage of affected land area and the percentage of affected population.

14. On Graph 1, plot your results for the percent of population affected (on the vertical axis) versus your results for the percent of land area affected (on the horizontal axis). Connect your points with a smooth curve that passes through the origin of the graph (0% land affected, 0% of the population affected).

Graph 1—Effects of sea level rise on population and cropland

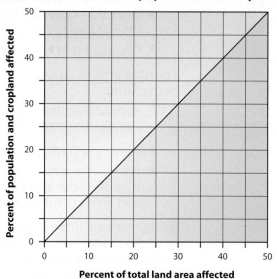

15. Plot the percentage of cropland affected versus the percentage of total land area affected on the same graph, using a dashed line.

The diagonal line divides the graph into two regions that show the relative strength of the effect of a rising sea level. Points plotted above the line (in the blue region) reflect a greater impact on people and cropland. Points plotted below the line (in the green region) reflect a greater impact on total land area.

16. Which is affected more strongly by a rising sea level, population or land area? What does this tell you about where people tend to live—near coastlines or in the interiors of continents? Explain.

Rising water and global population

The scenarios in this activity are extreme. A global rise in sea level of 1 to 2 meters is a more realistic hazard. You can use your data to construct a graph that will help estimate the effects of such a scenario.

17. On Graph 2, plot the percent of the world's total population affected versus the sea level rise. Connect your points with a smooth curve that passes through the origin of the graph.

Graph 2—Percent of population affected versus sea level rise

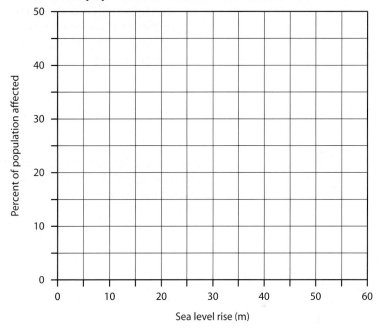

18. What percent of the world's population would be affected by a 2-meter rise in sea level? How about a 5-meter rise?

19. Are these scenarios of flooding a cause for concern? Explain.

Assessing the accuracy of your calculation

At the beginning of this activity, you estimated the rise in sea level due to the melting of the continental ice caps. One of the assumptions you made was that the change in the oceans' surface area was small enough to ignore. Later, however, you saw that the amount of land covered by the rising sea was quite substantial. In fact, in the 60-meter rise scenario, the amount of land covered by floodwaters is about half the size of the Atlantic Ocean. You can use this knowledge to refine your estimate of the rise in sea level.

20. What is the combined area of the oceans and the 60-meter flood-waters? (Refer to question 2b and Table 1 of this activity.)

    a. Surface area of the oceans =

    b. 60-meter flooded area =

    c. Combined area = (a) + (b) =

21. Calculate the rise in sea level that would occur if the continental ice caps melted completely, by dividing the total volume of water in the ice caps (question 2a in this activity) by the combined area (question 20c). Multiply the result by 1000 to convert from kilometers to meters, and round to the nearest meter.

This estimate for the rise in sea level is a lower limit; your original estimate is an upper limit. The actual rise in sea level would be somewhere between the two.

Activity 1.5

Comparing major reservoirs

In this unit, you have examined global water reservoirs, including lakes, rivers, ground water, the atmosphere, seas and oceans, and ice caps. Nearly all of the fresh water used around the world currently comes from surface and ground water reservoirs. Water from these reservoirs has been plentiful and relatively easy and inexpensive to obtain. However, as the global population grows, the demands on this resource are increasing. In many areas, the quality of water is declining as pollution contaminates the water supply with toxic chemicals and water-borne diseases. Providing an adequate supply of safe, usable water is becoming increasingly difficult and expensive. In the future, we will need to utilize other reservoirs to meet global demand.

Water in the atmosphere

When you consider the atmosphere as a water reservoir, be sure to include both water vapor and precipitation.

1. In Table 1 on page 27, discuss the advantages and disadvantages of using the oceans, atmosphere, and ice caps as sources of fresh water, according to each of the five listed criteria. (Use the back of the page if you need additional room.)

2. Which reservoir is currently not being used as a source of fresh water? Explain why not.

3. For each of the global water reservoirs below, list the characteristics of countries that would be most likely to utilize that reservoir as a source of fresh water.

   a. atmosphere

   b. oceans

   c. ice caps

The World Factbook

The World Factbook can be found at the CIA web site at:

http://www.cia.gov/cia/ publications/factbook

4. List the name of one country that you think best meets the criteria you listed in question 3 for utilizing each reservoir. Briefly explain why you think that reservoir is that country's best option for supplying their fresh water needs. (You may use the data in the **gwater.apr** ArcView project file along with an atlas, the World Factbook, or other reference sources to answer this question,)

    a. atmosphere

    b. ocean

    c. ice cap

Table 1—Comparison of global water reservoirs

Reservoir	Oceans	Atmosphere	Ice Caps
Accessibility and Delivery			
Quality			
Cost			
Sustainability			
Environmental Impact			

Unit 2
The Renewable Resource

In this unit, you will...

- *discover the importance of precipitation to our water supply,*

- *examine global and regional precipitation patterns, and*

- *investigate the factors that determine where precipitation falls and how it moves when it reaches land.*

Bonneville Dam spans the Columbia River 65 kilometers upstream from Portland, Oregon.

Activity 2.1

What is weather?

Weather is the state of the atmosphere at a given time and place, including temperature, moisture, wind, and pressure.

"Climate is what we expect...and weather is what we get."
Robert Heinlein

Too little, too much

In the previous unit you examined global water reservoirs and estimated the volume of water each contains. In this unit you will investigate the ways in which water moves between these global reservoirs, focusing on the two most visible manifestations of the hydrologic cycle, precipitation and runoff. Precipitation replenishes surface, soil and ground water resources critical to human society, while runoff in streams and rivers provides power and transportation and distributes water resources from regions with abundant water to regions where water is scarce.

Because precipitation and runoff can both be extremely variable when examined over the short term (days to months), this unit will focus on investigating long-term (30-year) precipitation and runoff patterns. You will identify factors that influence when and where precipitation occurs in the United States, and what happens to that precipitation after it reaches Earth's surface. Short-term precipitation and runoff events such as storms and floods can have dramatic economic and societal consequences, so you will begin this unit by considering major U.S. weather disasters.

Too little water

The Great Depression started with the crash of the stock market in 1929. By 1933, 25–30% of the workforce was unemployed and the stock market had lost 80% of its value. To recover from economic losses suffered during the peak years of the Depression, farmers plowed more land, planted more crops, and grazed their pastures more intensively, causing extensive soil damage. A prolonged drought caused a series of dust storms, or *black blizzards* to sweep across the central plains. The winds removed millions of tons of dry, damaged topsoil from barren fields and darkened skies in towns across the country. As a result, vast stretches of farmland were abandoned as farmers went bankrupt. The region hit hardest during this period was given the name *the Dust Bowl*.

To appreciate the magnitude of this event, read the eyewitness accounts on pages 35–37. The first story describes the dust storm of April 14, 1935, known as *Black Sunday*. The second account is an excerpt from the diary of Ann Marie Low, describing life on a North Dakota farm during the Dust Bowl years.

The Dust Bowl was a significant event, but not unique. Major droughts continue to occur in the Midwest. The story on page 38 describes the effects of a recent drought that struck Kansas during the summer of 2002.

Too much water

When the amount of precipitation falling on an area exceeds absorption, flooding can occur. Floods often hit areas downstream from where heavy precipitation occurs. The story on pages 39–40 tells about the 1993 flooding of the Mississippi River.

Billion dollar disasters

Most of the natural disasters to strike the United States are the result of having too little or too much water. The map below shows weather-related disasters in the U.S. between 1980 and 2001 that resulted in damage costing a billion dollars or more.

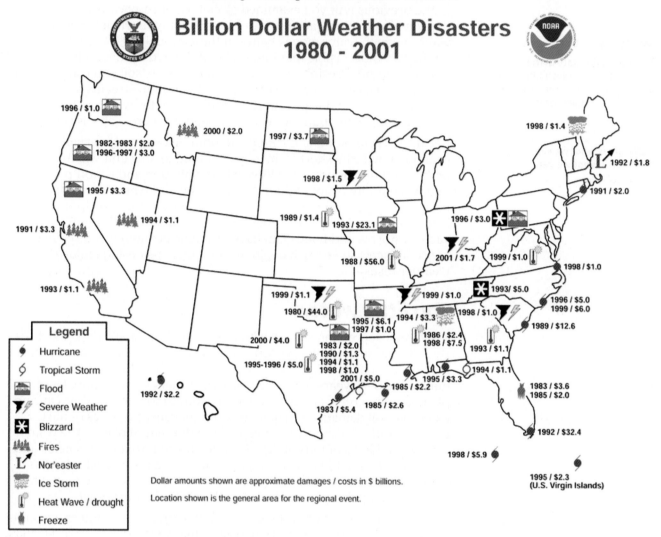

Questions

1. What types of weather disasters can be linked to too little water?

2. What types of weather disasters can be linked to too much water?

3. On the map, locate the four most costly weather disasters. Record the year, region, type of disaster, and cost in Table 1 below.

Table 1—Most costly U.S. weather disasters, 1980–2001

Year	Location	Type	Cost ($ billions)

4. Even if these disasters did not occur near you, how could they have affected you or your family?

5. Based on the reading on pages 36-37, how was Anne Marie Low's daily life affected by the drought?

6. As noted in the weather disaster map on page 32, the 1993 flood of the Mississippi River resulted in $23 billion in damage. After reading the story on pages 39–40, what factor do you think contributed the most to these monetary losses?

7. How would your daily life be affected by a major drought or flood today?

Eyewitness accounts

The Black Sunday dust storm

'Worst' Duster Whips Across Panhandle
Norther Strikes Sunday to Blot out Sun, Turn Day into Night

Amarillo, Texas, April 15 (AP)—North winds whipped dust of the drought area to a new fury Sunday and old timers said the storm was the worst they'd seen. A black duster—sun blotting cloud banks—raced over Southwest Kansas, the Texas and Oklahoma Panhandles, and foggy haze spread about other parts of the southwest.

A gentle, north breeze preceded 8,000-feet-high clouds of dust. As the midnight fog arrived, the streets were practically deserted. However, hundreds of people stood before their homes to watch the magnificent sight.

Arthur Rothstein, FSA-OWI Collection, Library of Congress

Dust storms often arose out of nowhere on otherwise beautiful days. Here, a car seems to be trying to outrun an approaching dust storm.

The storm struck just before early twilight. All traffic was blocked and taxi companies reported that it was difficult to make calls for nearly 45 minutes. Street signal lights were invisible a few paces away. Lights in 10 and 12 story buildings could not be seen.

John L. McCarty, editor of the Dalhart Texan, of Dalhart, the center of the drought-stricken area of the Panhandle, called a few minutes before the storm arrived in Amarillo. "I went outside the house during the storm and could not see a lighted window of the house three feet away." Mr. McCarty said.

NOAA Historic NWS Collection

A wall of dust approaches a Kansas town in 1935. Dust clouds often reached speeds of up to 60 miles per hour and were a mile or more high.

Darkness settled swiftly after the city had been enveloped in the stinking, stinging dust, carried by a 50-mile-an-hour wind. Despite closed windows and doors, the silt crept into buildings to deposit a dingy, gray film. Within two hours the dust was a quarter of an inch in thickness in homes and stores.

Damage to the wheat crop, already half ruined by drought and wind, could not be learned last night, but several grain men believed that the dust would cover even more of the crops.

The storm started yesterday when a high pressure area moved out of the Dakotas toward Wyoming, according to [*Amarillo Weatherman*] Mr. Collman. Most of the dust was from western Kansas and Oklahoma, he said.

NOAA Historic NWS Collection

Left-hand photo was taken in 1935 in Garden City, Kansas at 5:15 PM. Right-hand photo, taken 15 minutes later, shows the city during a dust storm that turned day into night, blotting out the sun. Note the street lights for orientation of the two photos.

Living in the Dust Bowl

The following excerpts are taken from the book Dust Bowl Diary *by Ann Marie Low.*

April 25, 1934, Wednesday

Last weekend was the worst dust storm we ever had. We've been having quite a bit of blowing dirt every year since the drouth *[drought]* started, not only here, but all over the Great Plains. Many days this spring the air is just full of dirt coming, literally, for hundreds of miles. It sifts into everything. After we wash the dishes and put them away, so much dust sifts into the cupboards we must wash them again before the next meal. Clothes in the closets are covered with dust.

Last weekend no one was taking an automobile out for fear of ruining the motor. I rode Roany *[her horse]* to Frank's place to return a gear. To find my way I had to ride right beside the fence, scarcely able to see from one fence post to the next.

Newspapers say the deaths of many babies and old people are attributed to breathing in so much dirt.

May 21, 1934, Monday

Saturday Dad, Bud, and I planted an acre of potatoes. There was so much dirt in the air I couldn't see Bud only a few feet in front of me. Even the air in the house was just a haze. In the evening the wind died down, and Cap came to take me to the movie. We joked about how hard it is to get cleaned up enough to go anywhere.

The newspapers report that on May 10 there was such a strong wind the experts in Chicago estimated 12,000,000 tons of Plains soil was dumped on that city. By the next day the sun was obscured in Washington, D.C., and ships 300 miles out at sea reported dust settling on their decks.

May 30, 1934, Wednesday

The mess was incredible! Dirt had blown into the house all week and lay inches deep on everything. Every towel and curtain was just black. There wasn't a clean dish or cooking utensil. There was no food.

It took until 10 o'clock to wash all the dirty dishes. That's not wiping them—just washing them. The cupboards had to be washed out to have a clean place to put them.

Mama couldn't make bread until I carried water to wash the bread mixer. I couldn't churn until the churn was washed and scalded. We just couldn't do anything until something was washed first. Every room had to have dirt almost shoveled out of it before we could wash floors and furniture.

We had no time to wash clothes, but it was necessary. I had to wash out the boiler, wash tubs, and the washing machine before we could use them. Then every towel, curtain, piece of bedding, and garment had to be taken outdoors to have as much dust as possible shaken out before washing. The cistern is dry, so I had to carry all the water we needed from the well.

That evening Cap came to take me to the movie, as usual. I'm sorry I snapped at Cap. It isn't his fault, or anyone's fault, but I was tired and cross. Life in what the newspapers call "the Dust Bowl" is becoming a gritty nightmare.

August 1, 1934, Wednesday

The drouth and dust storms are something fierce. As far as one can see are brown pastures and fields which, in the wind, just rise up and fill the air with dirt. It tortures animals and humans, makes housekeeping an everlasting drudgery, and ruins machinery.

The crops are long since ruined. In the spring wheat section of the U.S., a crop of 12 million bushels is expected instead of the usual 170 million. We have had such drouth for five years all subsoil moisture is gone. Fifteen feet down the ground is dry as dust. Trees are dying by the thousands. Cattle and horses are dying, some from starvation and some from dirt they eat on the grass.

October 1, 1934, Monday

Woodger, the federal acquisition agent, finally got around to see Dad last Saturday. From the kitchen I could hear the whole conversation, and it amused me. Dad was sitting on the back steps, resting after the noon meal, when a government car drove up. In this country, when anyone drives in, you meet

A farmer and his sons walk through a dust storm in Cimarron County, Oklahoma in April 1936.

him at the gate with hand outstretched, making him welcome. Dad knew this was the agent who had been dealing with the banks to rob our neighbors of their land. He didn't get up.

Woodger, a small man with a toothbrush-shaped mustache, walked up to the steps and introduced himself. He told about the proposed refuge and said he was there to appraise Dad's land and make on offer for it.

"It is not for sale."

In a sneering and condescending tone, "Oh, I believe this whole area is. All your neighbors are selling."

"No, they aren't. The banks are selling their places from under them. This place is not mortgaged."

Woodger spoke of what a great benefit this wildlife refuge will be—something for the good of all the people.

"I didn't build this ranch up for the benefit of all the people, but for me and my family."

Woodger pulled out every argument he could think of, then finally said, After all, this is submarginal land on which you can't make a living. "

That was news to Dad. He stood up, very slowly. The little pipsqueak agent stared in amazement as the bulk of Dad loomed above him. Dad is six feet four inches in his stocking feet, and higher with his boots on. He weighs 250 pounds, mostly bone and muscle. He is so big-boned and broad-shouldered it takes 250 pounds to flesh him out properly. Because his face has kept firm flesh and his hair has stayed jet black, he looks far younger than he is. His blue eyes, startling in his swarthy face, have never lost their keenness. By the time he had drawn himself up to his full height, the agent was open-mouthed.

Sand and dust buried farms and equipment, killed livestock, and caused human suffering during the Dust Bowl years.

"Young man, I want to tell you something. I've been here since the Territorial days. I started out with the clothes on my back and a $10.00 gold piece. I was young and dumb and uneducated. I didn't know I couldn't make a living here, and I didn't have any government expert to tell me so.

Dust cloud approaches Stratford, Texas on April 18, 1935.

"Young man, I've been fighting drouth and depression and blizzard and blackleg ever since the Territorial days. Everything you can see from here to the horizon belongs to me—the land, cattle, buildings, horses, and machinery. It is too late for you to tell me I can't make a living here. You better go away before you make me mad."

As Woodger scuttled for his car, Dad called after him, "By the way, when you get back to Washington, D.C., you can tell Franklin Delano Roosevelt I still have that $10.00 gold piece, too!"

Having gold is illegal now.

If Dad can get a decent price, he probably ought to sell. He is getting too old for a spread like this. Bud doesn't want it. Mama doesn't like it here. I love it, but am not going to. Everything I loved will be gone.

[Two years later…]

August 1, 1936, Saturday

July has gone, and still no rain. This is the worst summer yet. The fields are nothing but grasshoppers and dried-up Russian thistle *[tumbleweeds]*. The hills are burned to nothing but rocks and dry ground. The meadows have no grass except in former slough holes, and that has to be raked and stacked as soon as cut, or it blows away in these hot winds. There is one dust storm after another. It is the most disheartening situation I have seen yet. Livestock and humans are really suffering. I don't know how we keep going.

The dirt quit blowing today, so I cleaned the house. What a mess! The same old business of scrubbing floors in all nine rooms, washing all the woodwork and windows, washing the bedding, curtains, and towels, taking all the rugs and sofa pillows out to beat the dust out of them, cleaning closets and cupboards, dusting all the books and furniture, washing the mirrors and every dish and cooking utensil. Cleaning up after dust storms has gone on year after year now. I'm getting awfully tired of it. The dust will probably blow again tomorrow.

The Drought of June 2002

Kansas drought devastates wheat crop, forces widespread liquidation of cattle

By Roxana Hegeman

ELKHART, Kan. (AP) - Warren Bowker's combine kicks up a cloud of dust as he runs it nearly full speed across his thin stands of winter wheat. The machine almost touches the parched ground as it tries to cut stunted wheat that grew only a few inches tall.

Bowker's brother, Shaun, waves him in. Moments later they stare glumly at the combine's flat tire. Shaun Bowker uses his cell phone to call a repair shop, which says someone will be out soon. After all, there isn't much business these days.

It's been nearly a year since much of western Kansas has gotten substantial rain of even up to an inch, and the southwest corner has been hardest hit.

The drought has devastated the wheat crop now being harvested and spurred widespread selling off of cattle herds, as farmers become increasingly desperate to find enough feed and water to carry them through the summer grazing season.

Rural farm economies are hurting and even the wildlife is struggling to survive. The Bowker brothers are thankful to have anything left at all to harvest.

Poor crops are the least of their worries in this drought. Before the end of the month, the Bowkers will round up their cattle out of the Cimarron National Grassland and ship them off for sale, liquidating in one day what took them 10 years to build.

"We are going to dump the whole thing. We aren't going to fight it," Warren Bowker said.

Last week, forestry officials ordered all 100 farmers with permits to graze government lands to remove their grazing cattle from the drought-stressed grass. Usually 5,000 cattle feed off the national grassland; 3,200 are on it now, and all must leave before the end of June.

"The grass and vegetation is stressed so severely that to graze it will be detrimental," says Cimarron National Grassland district manager Joe Hartman.

Weather records dating back to 1913 show that never has there been less precipitation here than now. Even the Dust Bowl days of the 1930s logged more rain than this year, says Morton County Extension agent Tim Jones.

The big, black dust clouds of that era haven't repeated because much of the land has been put into the Conservation Reserve Program, a government program that pays farmers not to plant their cropland. But at times, big drifts of dirt blow across state highways so thickly that for a moment it seems like dusk. The drifting soil piles up along fence rows.

Activity at the Elkhart Co-op grain elevator - or lack of it - illustrates the troubles.

Manager Larry Dunn says his seven elevators usually take in 3.2 million bushels of wheat during harvest. This year they hope to collect 500,000 to 600,000 bushels. He figures 70 percent of the planted acres were abandoned long before harvest began.

"It is to the point it can get easily depressing for employees who have to hear it all the time," Dunn says.

Roughly 2,800 Kansas farmers have filed insurance claims for this year's crop, collecting $24 million so far even as losses mount with the start of the harvest, according to figures compiled by the federal Department of Agriculture's risk management agency.

Those figures only reflect claims paid, and the agency has a backlog. They don't include damage from a recent weekend hail storm that caused an estimated $6 million in damages to wheat crops.

This year's wheat crop in Kansas is insured for $645 million, and the money paid out so far is mostly for abandoned acres, says Rebecca Davis, the agency's director of the Topeka regional office.

"We are all trying to stay optimistic, but it is kind of bleak," says Pam Pate of Ben Pate Agency in Elkhart, noting that about 75 percent of the farmers who bought insurance from the agency have already filed claims for abandoned acres.

"It will turn around and get good again," she says. "We are hoping prices will come back up. It will rain or snow again. We are tough out here. We survive."

Elkhart has been through droughts before. Businesses come and go, but it will be mainly farmers who are forced to quit.

"A lot of our customers have no wheat left to cut," says Tim Predmore, service manager at a John Deere farm equipment dealership. "As far as we are concerned, there is no harvest."

The 1993 Mississippi Flood

In 1993, only three years after a severe drought caused billions of dollars in agricultural and industrial damage, a major flood occurred on the Upper Mississippi and lower Missouri Rivers. So widespread were the affects of this flood that it   has come to be called "The Great Flood."

Small Town Paid Its Dues Like Others Along The River

Mike Harden; Dispatch Columnist

NUTWOOD, Ill., July 21, 1993 (The Columbus Dispatch)— With his boots firmly planted on the half-submerged road outside Joyce's Kountry Korner, 1st. Lt. Steve Carney had the wary look of a man half-expecting a sucker punch.

But another wallop from the brown swirl of debris-laden flood water was hardly necessary. By yesterday, there was little left to do but tally the losses and clean up the mess.

Most of the nation was contemplating St. Louis' battle to save its south side from the surge of the Mississippi when little Nutwood went under. Just up the mouth of the Illinois River from the Gateway Arch, the tree-nestled village of 100 or so, wasn't much of a magnet for the TV crews and their satellite dish-topped trucks. But its fight was no less gallant than that in St. Louis.

Side-by-side with townsfolk, Illinois National Guard company commander Carney and the troops of Bravo Company, 133rd Signal Battalion, threw everything they had at the gorged Illinois River.

As late as Sunday afternoon, it appeared they might save Nutwood. The river, after all, would have to breach two levees and cut its way through 10,000 acres of prime farmland in

Residents and volunteers fill sandbags to stop further damage by rising flood waters.

the Nutwood Drainage and Levee District to even touch the tiny hamlet.

Carney recalls the instant he knew it had done just that. With water lapping over the two main levees early Sunday, he ordered his troops to climb on the bulldozers and push a dirt parapet up around the sandbagged outskirts of Nutwood to provide yet one more obstacle in case the worst happened.

It had, he knew, when he peered through his binoculars toward the main levee and saw two figures running up the road toward the town. "There was a gap 150 yards long in the levee," he said.

Moving like a surge of lava, the water was chasing the pair toward Nutwood. Carney jumped in his vehicle to rescue them, but a National Guard helicopter swooped in and plucked them to safety.

"Down the left-hand side of the road," Carney recalled, "it looked like the Niagara."

"It sounded like a train."

It was clear that the hastily erected parapet would only slow the inevitable.

'We were trying to buy time for the people to get their stuff out and evacuate," he continued. "We told them, 'Get your pictures and your family Bibles and enough clothes to stay a couple days."

Miles away, at the Jersey County Fair, a call went out over the public-address system, appealing for any available vehicles to aid in moving household possessions out of the stricken hamlet.

Off in the distance, up the dusty, sinuous road that winds down into Nutwood from nearby Fieldon, a long string of

An eerie calm settles on the water in this neighborhood in East Grand Forks, Minnesota. Many homes floated off their foundations, and all were significantly damaged.

An aerial view shows the extent of the flood damage.

pickup trucks - cavalry reinforcements - poured in from the county fair to help.

The water surged through the corn rows and into houses and mobile homes.

With all the people out and safe, the National Guard took care of one last, small lifesaving job: It unleashed two dogs, forgotten in the panic and still chained in yards.

Just around the bend, on a farm outside town, 78-year-old Harold Klunk made a last-ditch effort to rescue more than 3,000 bushels of corn that were stored in bins directly in the water's path.

He was aided in the frantic task by his nephew, farmer Jim Vahle, and Jim's sons, Matt and Gary.

With one of the last of the grain trucks loaded as the Monday sun was dipping toward dusk, a spry Klunk pulled himself up into the driver's seat and checked the rear view mirror.

Homes, businesses, and personal property were all destroyed by the high flood levels. These homes survived with little damage—others were not as fortunate.

Behind him lay 10,000 acres of submerged crops - his and all of his neighbors'. In the 75 years that had passed since he had been brought to the farm that had once been his father's, he had seen nothing like it.

In the yard, Vahle's wife, Wanda, fetched cold drinks for her husband and sons and watched, mildly amused, as mice and small rats scampered across floating twigs, trying to make it from submerged sheds to dry ground.

Bill Halemeyer and Lee Gettings, two of the neighboring farmers who helped Klunk save his grain, could legitimately say they had nothing better to do. Their own cash crop lay off in the distance toward Nutwood - under 20 feet of water.

Klunk fired up the truck and headed for the grain elevator.

Like the drought that struck the region in 1990, the floods caused widespread agricultural damage. This corn crop was a total loss.

The corn he was carrying would be just about the last grain carried to the elevator this year from the fertile Nutwood Drainage and Levee District.

And it was 1992's crop.

Not far away, Carney, who had been fighting the flood up in Quincy, Ill., before picking up the battle in Nutwood, awaited his next assignment.

"We're going to fight it all the way to Cairo (Ill.)," he gamely said, "but we're getting tired of losing."

Dusk was only minutes away when he gazed off to the west across the flood water toward the bluffs separating Illinois and Missouri.

"It looks like a beautiful sunset on a lake," he mused to no one in particular, "but there are people's homes under there."

When you have finished reading these eyewitness accounts, remember to answer the questions on page 32-33.

Activity 2.2

Global precipitation patterns

The atmosphere is one of the smallest global water reservoirs, yet it is one of the most important sources of water. Precipitation, in the form of rain or snow, provides the water that nourishes plants and animals, feeds our rivers and lakes, and replenishes the ground water stored in aquifers. Atmospheric water is the most easily renewed reservoir— a water molecule resides in the atmosphere an average of nine days before returning to Earth's surface. In this activity, you will investigate global and regional precipitation patterns, surface water runoff, evaporation, and transpiration to understand where precipitation falls and how it flows.

Precipitation and latitude

Aspects of weather, such as the amount and distribution of precipitation, vary considerably over short time periods. By averaging precipitation data over many decades, climate patterns appear at global, regional, and local scales. In this activity, you will explore these patterns.

▶ Launch ArcView and locate and open the **renewable.apr** file.

▶ Open the **Global Precipitation Patterns** view.

This view shows the annual global precipitation in centimeters per year.

1. Based on the global precipitation theme legend, what is the range of average annual precipitation on Earth? Be sure to include the units of measure.

2. Near what latitude is the average precipitation the highest?

3. In which latitude band in both the Northern and Southern Hemisphere (0°–30°, 30°–60°, 60°–90°) is the average precipitation the lowest?

▶ Turn on the **Prime Meridian** theme.

The prime meridian is another name for the 0° longitude line. Examine the precipitation patterns along the prime meridian, both north and south of the equator.

4. How does precipitation in the Northern Hemisphere vary with latitude along the prime meridian? Describe any patterns you see.

Latitude and longitude

Latitude lines run east and west. Latitude is the angular distance on Earth's surface measured north or south from the equator, ranging from 0° at the equator to 90° at the poles.

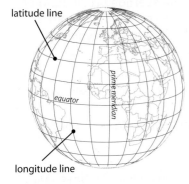

Longitude lines run north and south. Longitude is the angular distance measured east or west from the prime meridian, ranging from 0° at the prime meridian to 180° at the International Date Line.

To turn a theme on or off, click its checkbox in the Table of Contents.

5. How does precipitation in the Southern Hemisphere vary with latitude along the prime meridian? Are precipitation patterns the same in both hemispheres? If not, describe any differences you see.

Patterns of global precipitation are caused by uneven solar heating of Earth's surface. This heating is most intense at the equator, where evaporation and convection combine to lift large, moist air masses upward. The air masses cool as they rise, causing moisture to condense and clouds to form. In theory, the heaviest precipitation should be along the equator, because solar heating is highest there. However, the large continental land masses in the Northern Hemisphere increase heating north of the equator, shifting the band of precipitation northward.

Summarizing precipitation by latitude

Now you will quantify these global precipitation patterns. By creating a summary table, you will be able to determine the average precipitation for each 10-degree interval of latitude and graph the results. The summary table will combine data from the same latitudes in both hemispheres, helping to simplify the precipitation patterns.

To activate a theme, click on its name in the Table of Contents.

Help! My computer does not let me save files!

If you are using a computer that does not allow you to save files:

- close the **Global Precipitation** theme table.
- click the Media Viewer button 🖼 ;
- open the **Precipitation and Latitude** file from the media list;
- skip ahead to question 7.

▸ Activate the **Precipitation (cm/yr)** theme.

▸ Click the Open Theme Table button 🖩 to open the **Precipitation (cm/yr)** theme table.

▸ Click the **Latitude Range** field heading. The field name will be highlighted gray to show that it is selected.

▸ Click the Summarize button 🖻.

▸ In the Summary Table Definition dialog box, click the **Save As...** button. Navigate to the location where you wish to save the file, enter **Lat_Precip.dbf** as the file name, and click Save . (See note at left if your computer does not allow you to save files.)

▸ Set the **Field** to **Precipitation (cm/yr)** and the **Summarize by** operation to **Average**, then click the **Add** button. A new field called Ave_Precipitation should appear in the list box on the right.

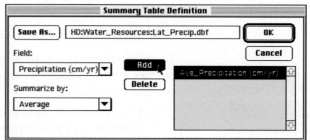

▸ Click **OK** to create the summary table. (This may take a minute, so be patient!)

The summary table should contain 9 rows, one for each 10-degree band of latitude from 0°–90°, and three columns: Latitude Range, Count, and Ave_Precipitation (cm/yr).

6. Complete the bar graph below using the average precipitation values you obtained from the summary table for each 10-degree band of latitude. The first latitude band has been plotted for you.

Graph 1—Precipitation (cm/yr) versus latitude

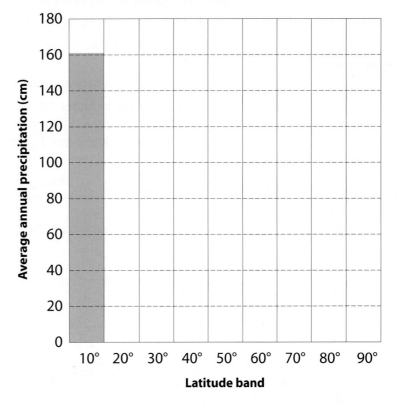

7. How does your graph of average precipitation by latitude compare to your earlier visual assessment of global precipitation in question 4?

▶ Close the summary table and the **Precipitation (cm/yr)** theme table windows, as well as the Precipitation and Latitude media window, if you opened it.

Precipitation in deserts and rainforests

Next, you will look at the relationship between precipitation, latitude, and the global distribution of two of Earth's important ecosystems, deserts and tropical rainforests.

▶ Turn on the **Deserts** and **Tropical Rainforests** themes.

To turn a theme on or off, click its checkbox in the Table of Contents.

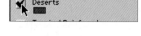

These themes show the locations of major world deserts and tropical rainforests.

8. On the globe diagram provided below:

- Mark the latitudes where deserts and tropical rainforests are found, using a different color or pattern for each ecosystem.
- Complete the legend by filling in the color or pattern for each type of ecosystem.

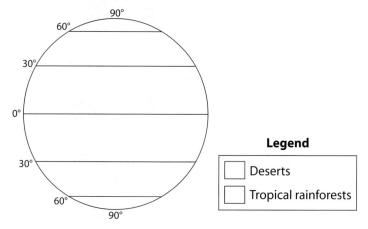

Legend

☐ Deserts
☐ Tropical rainforests

9. How do the locations of deserts and tropical rainforests compare to the graph of global precipitation by latitude on page 43?

Deserts form where air is descending, warming up, and drying out. Tropical rainforests, on the other hand, occur in areas where warm, moist air rises and cools. If the air cools sufficiently, the water vapor it contains condenses and falls as precipitation.

Precipitation in the deserts

You may be wondering, "How much more precipitation falls in the tropical rainforests than in the deserts?" Next, you will determine the average precipitation received by these ecosystems each year.

▶ Activate the **Precipitation (cm/yr)** theme.
▶ Choose **Theme ▶ Select By Theme**. Set the pop-up menus to read "Select features of the active themes that **Intersect** the selected features of **Deserts**," as shown at left.
▶ Click the **New Set** button.

Select By Theme settings

Set the Select By Theme dialog box as shown below.

The precipitation values that are within or touching the features of the **Deserts** theme will be highlighted orange.

To select a field, scroll across and click on the field name. Selected field names are highlighted gray.

▶ Click the Open Theme Table button 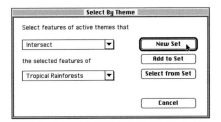 to open the **Precipitation (cm/yr)** theme table.

▶ Click the **Precipitation (cm/yr)** field heading. The field name will be highlighted gray to show that it is selected.

▶ Choose **Field ▶ Statistics...** to calculate statistics for the precipitation data.

10. Record the average (**Mean**), minimum (**Min**), and maximum (**Max**) precipitation values for the deserts in Table 1.

Table 1—Precipitation in desert and rainforest ecosystems

Region	Mean Precipitation (cm/yr)	Minimum Precipitation (cm/yr)	Maximum Precipitation (cm/yr)
Deserts			
Tropical Rainforests			
Global			

▶ Close the **Precipitation (cm/yr)** theme table.

11. According to Graph 1 on page 43, what other part of the world, in addition to the deserts, receives 29 cm of precipitation or less per year?

In Graph 1 on page 43, the precipitation at 30°N and 30°S is much higher than the estimate you calculated for the deserts occurring at these latitudes in Table 1. In fact, not all land at these latitudes is desert. You will examine other factors that contribute to the varied climate in these regions in the next several activities.

Precipitation in the tropical rainforest

▶ Activate the **Precipitation (cm/yr)** theme and choose **Theme ▶ Select By Theme.** Set the pop-up menus to read "Select features of the active themes that **Intersect** the selected features of **Tropical Rainforests**," as shown below.

▶ Click the **New Set** button.

The precipitation values that are within or touching the features of the **Tropical Rainforests** theme will be highlighted light green.

▶ Click the Open Theme Table button to open the **Precipitation (cm/yr)** theme table.

▶ Click the **Precipitation (cm/yr)** field heading. The field name will be highlighted gray to show that it is selected.

▶ Choose **Field ▶ Statistics...** to calculate statistics for the precipitation data.

12. Record the average (**Mean**), minimum (**Min**), and maximum (**Max**) precipitation values for the tropical rainforests in Table 1.

▶ Close the statistics window, but leave the **Precipitation (cm/yr)** theme table open. You will need it for the next section.

Average global precipitation

For comparison, determine the global average annual precipitation.

▶ Choose **Edit ▶ Select None** to deselect the rainforests.

▶ Click the **Precipitation (cm/yr)** field heading. The field name will be highlighted gray to show that it is selected.

▶ Choose **Field ▶ Statistics...** to calculate statistics for the precipitation data. (Be patient - this calculation may take a while.)

13. Record the Average (**Mean**) Global Precipitation in Table 1.

At left is a list of average annual precipitation figures for several U.S. cities.

14. In terms of average annual precipitation, which two cities in the U.S. are most similar to deserts? Which two cities are most similar to rainforests?

▶ Close the statistics window and the **Precipitation (cm/yr)** theme table.

▶ Close the **Global Precipitation Patterns** view.

Average annual precipitation for selected U.S. cities (cm/yr)

City	cm/yr
Baton Rouge, Louisiana	165.4
Boston, Massachusetts	111.3
Charlotte, North Carolina	109.7
Denver, Colorado	38.9
Hilo, Hawaii	327.4
Los Angeles, California	30.7
Nashville, Tennessee	123.2
New York City, New York	108.8
Salt Lake City, Utah	38.9
St. Louis, Missouri	86.1
Tucson, Arizona	28.2
Seattle, Washington	98.0

Activity 2.3
Moving air and water

Where the wind blows

Winds are caused by differences in temperature, which result in differences in air pressure. Where air is warm, it expands and rises, creating low pressure. Where air is cool, it contracts and sinks, creating high pressure. Generally, winds blow from areas of high pressure to areas of low pressure as the atmosphere acts to equalize the pressure differences.

Global surface winds

Uneven heating of Earth's surface by the sun and the force generated by Earth's rotation combine to create wind belts that encircle the planet. Near the equator, where the sun's rays strike the surface most directly, the air is heated, causing it to expand and rise. This forms a low-pressure belt around the equator. At the surface, cooler air north and south of the equator flows toward the low pressure belt to replace the rising air. The Coriolis effect, caused by Earth's rotation, deflects these winds to the right in the Northern Hemisphere and to the left in the Southern Hemisphere. Thus, rather than blowing directly toward the equator, the winds blow from east to west in both hemispheres. These easterly winds are called the **trade winds**.

The rising air at the equator condenses and precipitates much of its moisture, creating the tropical rainforests. As the air rises, it travels toward the poles, eventually cooling, drying out, and becoming more dense. At about 30° north and south latitude, this cool, dense air sinks, forming a band of high pressure that is responsible for the deserts found at these latitudes. When this sinking air reaches the surface, some of it spreads toward the poles. This air is again deflected by the Coriolis effect, but this time the winds blow from west to east. These winds are known as the **prevailing westerlies**.

At the poles, cold, sinking air creates a high pressure area. As the sinking air moves toward the equator, it is deflected toward the west, forming the wind belt called the **polar easterlies**. At around 60° latitude, this air warms and rises, forming a weak low pressure zone. This is where most winter storms begin.

Air masses and fronts

Most weather is caused by interactions between **air masses** in the lower atmosphere, which extends from Earth's surface to an altitude of about five kilometers. Air masses are large volumes of air that remain in one place long enough to acquire the humidity and temperature characteristics of the surface beneath them.

Continental air masses, abbreviated **c**, form over land and are generally dry. **Maritime** air masses, abbreviated **m**, form over oceans and are generally moist.

Tropical air masses, abbreviated **T**, form near the equator and are typically warm. **Polar** air masses, abbreviated **P**, form at high latitudes and are cold. The coldest air masses form near the poles, and are called **arctic** air masses, abbreviated **A**.

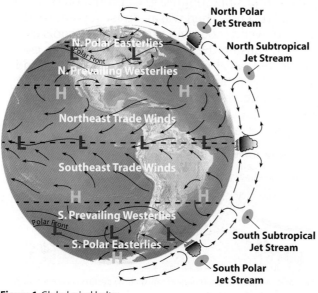

Figure 1. Global wind belts.

Combining these characteristics defines six main types of air masses that affect North America, as shown in Figure 2.

- **cT**—continental tropical
- **cP**—continental polar
- **cA**—continental arctic
- **mT**—maritime tropical
- **mP**—maritime polar
- **mA**—maritime arctic

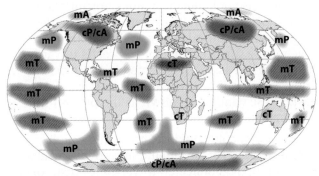

Figure 2. Global air masses.

The boundaries between air masses are called **fronts**. The temperature and humidity differences along these frontal boundaries produce clouds, precipitation, wind, and other weather phenomena. The three basic types of fronts—cold fronts, warm fronts, and stationary fronts—are illustrated in Figure 3.

Precipitation, air masses, and topography

To get rain or snow, two conditions must be met. First, the lower atmosphere (0-10,000 feet) must have large quantities of water vapor, or **precipitable water**. Second, the air mass must cool sufficiently. Cooling occurs when an air mass is forced upward. There are four types of uplift:

- **convective uplift**, where an air mass moves over a warmer surface, expands, and rises.
- **frontal uplift**, where a denser, cooler air mass "wedges" underneath a warmer air-mass and forces it upward (Figure 3);
- **orographic uplift**, where an air mass is forced up and over a landform such as a mountain or high plateau (Figure 4);
- **disturbance uplift**, where conditions in the upper atmosphere, often associated with the jet stream, produce uplift in the lower atmosphere.

Figure 3. Three basic types of weather fronts

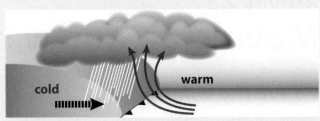

In a **cold front**, cold air displaces warm air along a steeply-sloped boundary. The heavier cold air wedges under the warm air, pushing it upward. Precipitation is often heavy and occurs behind the front.

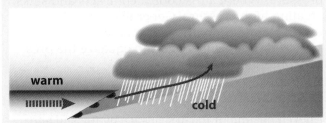

In a **warm front**, warm air replaces cooler air. The warm air slides over the heavier cold air, creating a boundary with a gentle slope. As the warm front approaches, thin, high clouds gradually grow thicker and lower. Warm front precipitation is usually steady and falls ahead of the front.

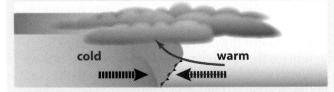

In a **stationary front**, the boundary between the air masses is not moving. Warm air continues to move over the dense, cooler air, causing clouds to form. Precipitation, if present, is typically slow and steady.

The rainshadow effect

Orographic uplift (Figure 4) is a common phenomenon. It occurs when warm, moist air is pushed up and over a mountain range. The air cools as it rises, causing the water vapor in the air to condense and fall as either rain or snow on the windward side of the mountains. Having lost much of its moisture, the air descends the leeward side of the mountain range, compresses, and warms, creating a warm, dry wind. Valleys and lowlands on

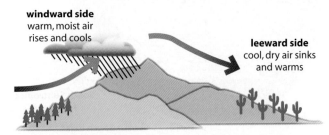

Figure 4. Orographic uplift creates rainshadow deserts on the leeward slopes of mountain ranges.

the leeward side receive much less precipitation and are said to be in the *rainshadow* of the mountains, in some cases forming **rainshadow deserts**.

In the Pacific Northwest, where maritime polar (mP) air masses from the northern Pacific Ocean carry moisture inland, the north-south running Cascade Mountains produce a strong rainshadow effect. While most people think of Washington and Oregon as wet and rainy, the eastern half of those states on the leeward side of the Cascades are deserts.

Upper level winds—the jet streams

During World War II, bomber pilots flying at altitudes above six kilometers confirmed the existence of upper-level air currents predicted by atmospheric scientists to blow at speeds of nearly 300 kilometers per hour. These high-altitude "rivers of air," called **jet streams**, play an important role in the formation and movement of storms.

Although television meteorologists usually refer to a single "jet stream," there are actually two main jet streams in each hemisphere. These are the **polar jet stream** and the **subtropical jet stream**, and they encircle the globe along winding paths thousands of kilometers long and hundreds of kilometers wide.

The polar jet stream flows at an altitude of 5-12 kilometers along the polar front—the boundary between cold, dry high-latitude air and warmer, moist middle-latitude air. The polar jet stream influences precipitation in the U.S. by contributing to the formation and movement of storms along the polar front.

The subtropical jet stream flows at an altitude of 8–15 kilometers above the boundary between cool mid-latitude air and warm tropical air. The subtropical jet stream contributes moisture to summer storms in the southern U.S., but is much weaker and has less influence on weather patterns than does the polar jet stream.

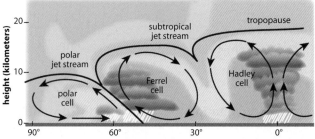

Figure 5. Cross-section of Earth's atmosphere, showing the locations of the jet streams.

Where the water flows

The interaction between air masses, global winds, and topography determine where precipitation occurs. When precipitation reaches the ground, gravity continues to pull the water downward. The characteristics of the surface influence the water's path. If the surface is **permeable**, permitting water to pass through, and **porous**, containing empty spaces, water can move through or **infiltrate** it.

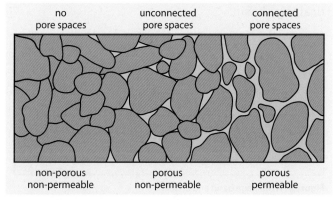

Figure 6. Illustration of how grain packing affects the porosity and permeability of an aquifer.

If Earth were flat and featureless, permeability and porosity would completely dictate water flow. However, Earth has mountains, hills, and valleys, so water flow is determined by the interaction between the surface properties and surface features, or **topography**.

Water in the soil can return directly back to the atmosphere through **evaporation**; it can be taken up by plant roots and returned to the atmosphere by **transpiration**; or it can continue infiltrating downward to join ground water in a water-bearing rock layer, or **aquifer**.

Topography: slope and aspect

Two facets of topography, slope and aspect, affect water flow. **Slope** is the steepness of the surface. Expressed as an angle, slope influences the *speed* that water flows downhill. For example, water moving down a mountain with a 40° slope flows faster than water flowing down a 10° slope. **Aspect** describes the direction that the slope faces—north, south, east, or west—and determines the *direction* that water flows on the surface.

Drainage basins, systems, and divides

Aspect controls the accumulation of runoff from precipitation. **Drainage basins** are areas in which runoff flows downhill to a common point, such as a lake, or to a common channel, such as a river or stream, as shown in Figure 7. Within the basin, the land aspect faces inward. A drainage basin is characterized by a **drainage system**, a network of interconnected creeks, streams, and rivers. The Mississippi River Basin is an example of a large drainage basin.

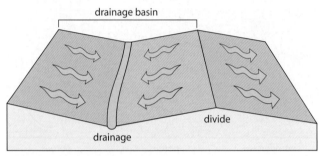

Figure 7. A drainage basin and divide.

Divides form the boundaries between drainage basins. At a divide, the opposing aspects of the land cause water to flow away from the boundary. The Rocky Mountains are an example of a divide. Precipitation that falls on the western slopes of the Rockies eventually drains into the Pacific Ocean or the Gulf of California. Precipitation that falls on the eastern slopes eventually drains into the Gulf of Mexico.

Watersheds

Drainage basins and divides define regions known as **watersheds**. A watershed is the total area drained by a river and its tributaries. Figure 8 shows the Colorado River watershed, a large, multi-state region that includes networks of creeks, streams, and rivers that drain into the Colorado River. The term watershed can also be applied to relatively small areas. For example, the land that drains into a small pond may also be called a watershed.

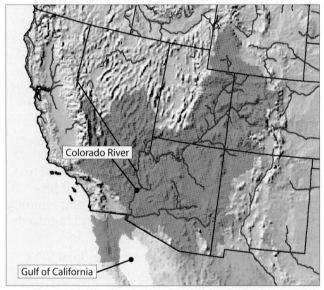

Figure 8. The shaded area represents the Colorado River watershed (or Basin). Water from this region eventually drains to the Colorado River and the Gulf of California.

Questions

1. How does the jet stream influence precipitation patterns?

2. Explain how air masses and topography interact to produce rainshadows.

3. What determines the movement of precipitation after it reaches the ground?

4. How do permeability and porosity affect the amount of precipitation that infiltrates the ground?

Activity 2.4

US precipitation patterns

In this activity, you will examine the precipitation patterns that characterize the climate of the United States. At this scale, other factors besides latitude become important in determining the distribution, quantity, and timing of precipitation.

East versus West—regional precipitation

First, you will explore precipitation patterns in the United States, paying particular attention to the relationship between topography and precipitation. Also, notice the role the oceans play in providing a source of atmospheric moisture that eventually precipitates over land.

topography—The physical features of a region; mountains, plains, valleys, etc.

▶ Launch ArcView and locate and open the **renewable.apr** file.

▶ Open the **US Precipitation Patterns** view.

This view shows the annual precipitation for the U.S. in centimeters per year. Notice how dramatically the precipitation varies across the country, particularly in the Western states.

1. Which regions of the U.S. receive the most precipitation?

▶ Zoom in on each of these regions that receive the most precipitation using the Zoom In 🔍, Zoom Out 🔍, and Pan tools ✋.

2. In areas with high precipitation, does precipitation increase or decrease with distance from the coastline?

To turn a theme on or off, click its checkbox in the Table of Contents.

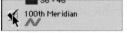

To activate a theme, click on its name in the Table of Contents.

West of the 100ᵗʰ meridian

When you select the area west of the 100th meridian (the purple line), try to select as close to the line as possible without going past it.

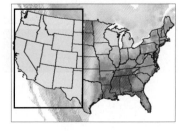

In 1879, the explorer and scientist John Wesley Powell noted that a natural dividing line based on precipitation seemed to separate the eastern and western U.S. This line is near the 100th meridian (100° west longitude). Next, you will determine how much these regions differ.

▶ Turn on the **100th meridian** theme. It appears as a heavy purple line that cuts the U.S. roughly in half.

▶ Scroll down the Table of Contents and activate the **US Precipitation** theme.

▶ Use the Select Features tool 🔲 to drag a box that encloses all of the area west of the 100th meridian (see sidebar). The selected region will be highlighted yellow.

▶ Click the Open Theme Table button 🔳 to open the **US Precipitation** theme table.

▶ Click the **Precipitation (cm/yr)** field heading. The field name will be highlighted gray to show that it is selected.

▶ Choose **Field ▶ Statistics...** to calculate statistics for the precipitation data. **This is a large dataset, so this may take time.**

3. Enter the mean (**Mean**), minimum (**Min**), and maximum (**Max**) precipitation values for the western U.S. in Table 1.

Table 1—Comparing precipitation east and west of the 100th meridian

	west of 100° W	east of 100° W
Mean Precipitation		
Minimum Precipitation		
Maximum Precipitation		

▸ Click the Switch Selection button 🔁 to select the region east of the 100th meridian.

▸ Repeat these steps to calculate precipitation statistics for the eastern U.S. and enter them in the table.

4. Using these values, compare the differences in mean, maximum, and minimum precipitation east and west of the 100th meridian.

▸ Close the statistics window and turn off the **100th meridian** theme.

▸ Choose **Theme ▸ Clear Selected Features** to deselect the eastern states.

Precipitation and air masses

After exploring some of the general patterns of precipitation across the U.S., it is time to consider some explanations for the patterns you have observed. For precipitation to occur, the lower atmosphere must have large quantities of water vapor, or precipitable water. The regions where air masses originate affect the amount of precipitable water they are likely to contain.

▸ Turn on and activate the **Air Mass Sources** theme.

▸ Click the Zoom to Full Extent button 🔲 to show the entire the air mass theme.

What is precipitable water?

Precipitable water is measured in terms of the depth of water that would be produced if all of the water vapor over an area were condensed to a liquid.

What is an air mass?

An air mass is a large volume of air with a fairly uniform temperature, pressure, and moisture content.

The colors of the **Air Mass Sources** theme represent the amount of precipitable water in each air mass. Examine the source regions (maritime or continental and arctic, polar, or tropical) and relate the amount of precipitable water to the source region's climate.

5. Use the color spectrum below to rank these five air mass types from driest to wettest: **mP**, **cP**, **mT**, **cA**, and **cT**. The first one has been done for you.

Driest   **Wettest**

cA

6. Which air masses hold more precipitable water—those that form over continents, or those forming over oceans?

7. Which air masses hold more precipitable water—those that form in tropical regions, in polar regions, or in arctic regions?

▶ Turn off the **US Precipitation** and **Air Mass Sources** themes.

Precipitation and topography—orographic uplift

When an air mass encounters mountains, it is pushed upward. If the air mass contains enough moisture and is cooled sufficiently, precipitation will occur. Now you will investigate how topography affects precipitation, both regionally and locally.

The **US Topography** theme is a shaded relief map of the United States, with major mountain ranges and basins visible as wrinkles in the map. Areas with little topographic variation appear smooth, almost featureless. Look at the patterns across the U.S. and answer the following question.

8. Compare the topography in the western U.S. to the topography of the eastern U.S.

Temperature and elevation

As air rises, the pressure decreases, causing the air to expand and cool. As air sinks, the pressure increases, compressing the air and causing its temperature to increase. The average rate of temperature change with changing elevation, called the *lapse rate*, is about 6.5°C/km.

Now you will look at topography and precipitation together, by making the precipitation theme semi-transparent.

▶ Double-click the **US Precipitation** theme in the Table of Contents to open the Legend Editor.

▶ Click the **Load...** button. In the dialog box that appears, locate and open the **Data** folder, select the file **transparent.avl**, and click **Open** (or **OK**).

▶ In the next dialog box, choose **Precipitation** from the **Field** pop-up menu, and click **OK**.

▶ Click **Apply** in the Legend Editor and close the Legend Editor window.

▶ Turn on the **US Precipitation** theme.

You should now be able to see through the **US Precipitation** theme to the underlying shaded relief image.

9. Describe the relationship between precipitation and topography. Compare the precipitation and topography of the Eastern U.S. to the Western U.S.

▶ Turn off the **Canada Topography** and **Mexico Topography** themes.

Precipitation in the Pacific Northwest

The Pacific Northwest includes the states of Washington and Oregon. This region is famous for its moist climate that results in lush rainforests.

10. Which type of air mass typically brings moisture to this region?

Understanding profiles

A **profile** is a side view of a surface produced along a line called a **transect**.

- A **topographic profile** shows the elevation of the land along a transect.

- A **precipitation profile** shows what it would look like if you piled up all of the precipitation along a transect for a year and viewed it from the side. The peaks are the places that received the most rainfall and the valleys are the places that received the least precipitation.

To use the Hot Link tool, position the *tip* of the lightning bolt cursor over the feature and click.

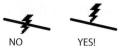

NO   YES!

Next, you will use a cross-section or profile of this region to explore the relationship between precipitation and elevation. A topographic profile will show local relief—mountain ranges and basins—as well as the overall elevation trend. A precipitation profile along the same transect will help you understand how precipitation relates to topography.

▶ Turn on and activate the **Profiles** theme.

This theme adds two transects, one that extends from the state of Washington east into Idaho, the other from Louisiana north into Illinois.

▶ Click on the Washington–Idaho transect using the Hot Link tool ⚡.

In the window that opens, the upper profile shows annual precipitation and the lower profile shows the elevation along the transect. You may want to arrange the windows on your screen so you can see both the transect line in the view and the window containing the profiles.

Using the profiles, answer these questions:

11. Which mountains receive the most precipitation—those near the coast, or those farther inland?

12. Within a given mountain range, how does the precipitation on the western slope of each mountain range compare to the precipitation on the eastern slope of the same range?

13. Explain this pattern in terms of orographic uplift.

▶ Close the **Washington–Idaho Profile** window.

Precipitation in the Southeast

Now, you will examine precipitation and topography patterns in the southeastern U.S.

14. Which type of air mass typically brings moisture to this region?

▶ Click on the Louisiana–Illinois transect using the Hot Link tool ⚡.

These profiles show the precipitation and topography along a transect from Louisiana to southern Illinois. For easy comparison, the vertical scales of the profiles are the same as those of the WA–ID profiles.

Use the profiles to answer these questions:

15. Describe the topography of the states crossed by the transect. (Are they mountainous, flat, hilly?)

16. Is the amount of precipitation along this transect influenced more by topography or by the distance from the coastline? Explain.

▸ Close the **Louisiana–Illinois Profile** window.
▸ Turn off the **US Precipitation** and **Profiles** themes.

17. Based on your observations, describe how the movement of air masses, distance from the ocean, and topography affect the amount of precipitation a location receives.

Precipitation and the polar jet stream– frontal and disturbance uplift

The position of the polar jet stream exerts a strong influence on precipitation patterns in the U.S. The polar jet stream flows above the polar front—the boundary between cold, dry air to the north and warm, moist air to the south—and contributes to the formation of frontal storms that bring substantial amounts of precipitation to the northern and central U.S., particularly during the winter.

▸ Click the Media Viewer button 🎞 and choose the **Jet Stream Movie** from the media list.

This animation shows the changing position of the jet stream over the course of a year, averaged from 30 years of data. Watch the movie several times, noting the position and speed of the jet stream at different times of the year. The jet stream always flows west to east, but its latitude varies with the seasons. The colors represent the wind speed, with reds and oranges representing the fastest winds.

18. During which season is the jet stream farthest north?

19. During which season is the jet stream farthest south?

Help! The movie does not play!

If the movie does not play, or does not play smoothly, you may need to install a newer version of QuickTime Player. For more information, see the **Troubleshooting** section on page *xiii*.

Seasonal jet stream views

If you have difficulty interpreting the animation, this view also includes themes showing the average position and wind speed of the jet stream for the months of January, April, July, and October. Keep in mind that this is an average position—the actual position of the jet stream is constantly changing.

20. During which season is the speed of the jet stream the fastest, and in which season is it slowest?

   a. fastest –

   b. slowest –

   ▶ Close the **Jet Stream Movie** window and return to the **US Precipitation Patterns** view.

   ▶ Turn off all themes *except* the **US State Borders** theme.

   ▶ Turn on the **Jet Stream Frequency** theme.

This theme combines four monthly average jet stream positions, one from each season of the year. Areas are ranked according to the number of seasons that the average jet stream position crosses that region, from 0 to 4.

21. On the map below, circle and shade in the region(s) where the jet stream passes over four times a year. Circle but *do not shade* the region(s) where the jet stream crosses only once a year.

   ▶ Turn off the **Jet Stream Frequency** theme and turn on the **US Precipitation** theme.

Examine the relationship between precipitation and the number of times the jet stream is positioned over a region.

22. In what areas do average annual precipitation and the jet stream position appear to be related? Discuss the examples you selected on the map above.

23. Most of the southeastern U.S. from Texas to Florida has the same jet stream frequency (2), yet the precipitation in this region varies considerably. Explain how other factors affect precipitation there.

   ▶ Turn off the **US Precipitation** theme.

Precipitation and thermal convection— convective uplift

The jet stream does not explain all precipitation in the U.S. The final piece of the precipitation puzzle is the thunderstorms that rumble and roll throughout much of the country each summer. Thunderstorms develop when intense heating at the surface causes thermal convection. The heated air rapidly moves upward, cools, and the water vapor precipitates out. Thunderstorms provide a substantial amount of the annual precipitation for some regions of the country.

▶ Turn on and activate the **Summer Thunderstorms** theme.

This theme shows the average number of days in which thunderstorms occurred during July, August, and September.

24. In which two parts of the country are summer thunderstorms most common?

25. Which type of air mass provides the moisture that feeds these thunderstorms?

Now you will take a closer look at one of these areas.

▶ Click the Query Builder button ◀ to open the Query Builder dialog box.

▶ Enter the expression **[State_name] = "Florida"** and click the **New Set** button. The state of Florida will be highlighted yellow.

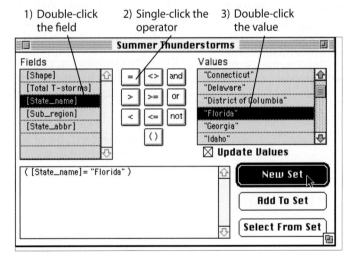

▶ Close the Query Builder dialog box.
▶ Click the Open Theme Table button ▦ to open the **Summer Thunderstorms** theme table.
▶ Click the **Days w/T-storms** field heading. The field name will be highlighted gray to show that it is selected.
▶ Choose **Field ▶ Statistics...** to determine the average number of thunderstorms Florida has each year, given as the **Mean** value.

Watch a summer thunderstorm!

To view time-lapse movies of summer thunderstorms, click the Media Viewer button 🔳 and choose **Thunderstorm Movie 1, 2,** or **3**.

26. What is the average number of thunderstorm days in Florida?

   ▸ Close the statistics window and the **Summer Thunderstorms** theme table.

27. What percentage of summer days do thunderstorms occur in Florida? (Hint: Divide the average number of thunderstorms by the number of days in July-September and multiply the result by 100.)

Now you will look at a southwestern state, New Mexico.

   ▸ Repeat the query process above using the expression **[State_name] = "New Mexico"**.

   ▸ Repeat the field statistics operation to get the average (**Mean**) number of thunderstorms for New Mexico during the summer.

28. What is the average number of thunderstorm days in New Mexico?

   ▸ Close the **Summer Thunderstorms** theme table.

29. What percentage of summer days do thunderstorms occur in New Mexico?

To determine how important thunderstorms are to the Southwest and Southeast, you will study a movie that shows the average percentage of the annual precipitation that each state receives in each month.

   ▸ Click the Media Viewer button 🖼 and choose **US Monthly Precipitation** from the media list. Play the movie several times and watch how precipitation varies across the country throughout the year.

30. What percentage of its annual rainfall does New Mexico receive from July through September? Use the highest percentage you see in New Mexico in the movie for your answer.

31. What percentage of its annual rainfall does Florida receive from July through September? Use the highest percentage you see in Florida in the movie for your answer.

32. Where do more summer thunderstorms occur—in the Southwest or in the Southeast? Where are summer thunderstorms more important to the annual precipitation—in the Southwest or in the Southeast? Explain your answer.

   ▸ Close the **US Monthly Precipitation** movie window.

How many days?

"Thirty days hath September...", July and August both have thirty-one, so the summer data cover a total of 92 days.

Help! The movie does not play!

If the movie does not play, or does not play smoothly, you may need to install a newer version of QuickTime Player. For more information, see the **Troubleshooting** section on page *xiii*.

Activity 2.5

Surface water flow

In this activity, you will examine what happens to precipitation when it reaches Earth's surface.

Aspect, divides, and watersheds

To better understand water flow, you will begin by looking at the *aspect*—the main direction in which the land slopes—of regions within the U.S. Water flows downhill along the steepest path, so aspect indicates the direction that runoff will flow on a surface.

▶ Launch ArcView and locate and open the **renewable.apr** file.

▶ Open the **Surface Water Runoff** view.

In this view, the **Land Aspect** theme uses different colors to show the general direction in which precipitation falling on a given land area flows, based on aspect. The U.S. has been divided into four regions—West, West Central, East Central, and East. You will use the **Land Aspect** theme to determine the primary direction that water flows in each of these regions. The graphic at left shows the colors used to represent the direction the slope is facing.

▶ Use the *predominant color* to determine the general direction that water flows in each region. (For example, if more of the region looks red than any other color, water flows toward the north.)

▶ If you need help determining the flow direction, click the Media Viewer button 📷 and choose the **US Land Aspect by Region** file. The graphs show the percent area by aspect for each region.

1. On the map below, draw arrows showing the primary direction of water flow in each of the four major regions of the U.S.

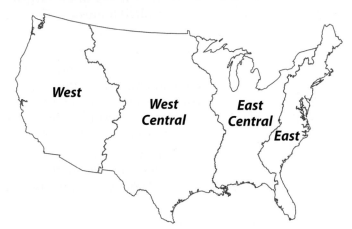

2. How many continental divides are evident from the flow direction arrows you put on the map? Mark the divide(s) on the map using heavy dashed lines.

Aspect directions

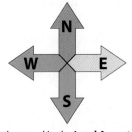

The colors used in the **Land Aspect** theme represent the general direction that runoff will flow in each region.

3. Shade in the large drainage basin where precipitation runoff flows toward a central area. (Hint: drainage basins are bounded by divides.)

4. On the map, mark the major river that drains this basin with a heavy solid line.

5. What major U.S. river system does this watershed form? (Hint: it is the largest river system in the U.S.)

▶ Turn off the **Land Aspect** and **Aspect Regions** themes.

To turn a theme on or off, click its checkbox in the Table of Contents.

Watersheds and river systems

▶ Turn on the **Major US Watersheds** theme.

The **Major US Watersheds** theme shows the locations of the major river basins in the United States. This part of the activity explores the six largest watersheds.

▶ To highlight the six largest watersheds:
- Activate the **Major US Watersheds** theme.
- Click the Open Theme Table button 🖼 to open the **Major US Watersheds** theme table.
- Click the **Watershed Area (km2)** field heading. The field name will be highlighted gray to show that it is selected.
- Click the Sort Descending button 📉 to sort the field from largest to smallest.
- Holding the shift key down, use the Pointer tool 🔺 to click and drag down the top six rows of the theme table. The rows should be highlighted yellow, indicating that they have been selected.
- Close the theme table and look at the view window. The six watersheds you selected should be highlighted yellow. If not, open the theme table and try the selection process again.

6. How many of the largest watersheds are completely within the United States?

Remember that a watershed is an area drained by a network of inter-connected streams and rivers. Runoff and smaller streams flow into a common body of water—usually a larger stream or river. Therefore, everyone living downstream in a watershed is affected by what happens upstream, in terms of the quality and quantity of the water that they use for drinking, irrigation, etc.

7. How might the fact that watersheds do not follow political borders complicate their management? Answer the question in terms of both water quality and water quantity.

Measuring stream discharge

Stream **discharge** is the volume of water moving past a point in a given length of time. To calculate discharge, you must know the depth and speed of the water. These are measured at a **gauging station** like the one shown below.

USDA/ARS/Southwest Watershed Research Center

Gauging station at the outlet of the USDA/ARS Walnut Gulch Experimental Watershed near Tombstone, Arizona. Shown here during a flash flood, this outlet drains an area covering 148 square kilometers.

Kilometers to centimeters

1 km = 1,000 m
1 m = 100 cm
So, 1 km = 100,000 cm

River discharge and basin runoff

In this section, you will explore how runoff varies by region in each of the six largest U.S. watersheds. It is not possible to directly measure runoff over an entire watershed, but you can measure the watershed's *discharge* and relate it to the runoff.

▶ Turn on the **Major US Rivers, Canadian Rivers, Mexican Rivers**, and **Gauging Stations** themes.

The **Gauging Stations** theme shows the locations of the gauging stations that measure discharge for each of the six major watersheds.

▶ Activate the **Major US Watersheds** theme.

▶ Click in one of the highlighted watersheds using the Identify tool **ⓘ** to find the name and data for that watershed.

8. Record the **Watershed Area (km²)** and **Annual Precipitation (cm)** for each watershed in columns 3 and 6 of Table 1 on page 63.

▶ Repeat this process to record the area and annual precipitation for the other five watersheds.

▶ Activate the **Gauging Stations** theme.

▶ Click directly on a gauging station symbol using the Identify tool **ⓘ** to see the data for that station.

9. Record the **Annual Discharge (km³)** of the main outflow river for each watershed in column 2 of Table 1.

▶ Repeat this process to record the annual discharge for the other five watersheds.

10. For each named watershed, calculate the runoff in kilometers of water per square kilometer of surface area. To do this, divide the value in column 2 by the value in column 3, and record the result in column 4 of Table 1.

11. Convert the runoff in column 4 from kilometers to centimeters by multiplying kilometers by 100,000. Record your results in centimeters in column 5 of Table 1.

Calculating infiltration and evapotranspiration

In Table 1, you calculated the amount of water that would have to run off every square meter of land in each watershed to produce the runoff recorded at the gauging station. Not every drop of precipitation simply flows over land and into a river. Precipitation can also evaporate directly from the ground, infiltrate the soil to be taken up by plant roots and transpired through the leaves, or infiltrate downward, eventually entering the watershed's ground water.

The mathematical equation describing runoff is:

Runoff = Precipitation – Evapotranspiration – Infiltration

Measuring evapotranspiration and infiltration are very difficult. Fortunately, you have information about precipitation and runoff. Therefore, you can determine the amount of precipitation lost through evaporation and infiltration by rearranging the equation.

Precipitation – Runoff = Evapotranspiration + Infiltration

12. Using the above equation, calculate the amount of precipitation that either infiltrates the ground or is lost through evapotranspiration. Record your results in column 7 of Table 1.

13. Calculate the percentage of precipitation that becomes runoff for each watershed and enter it in column 8 of Table 1.
    Percent Runoff = Runoff (column 5) ÷ Precipitation (column 6) × 100

14. Compare the average precipitation to the runoff you calculated for each watershed. Describe any patterns you see.

15. How does the percentage of precipitation that becomes runoff vary with climate shown in column 9? Why do you think this occurs?

16. Which of these major U.S. watersheds do you live in, or is nearest to you? How does the percent runoff you calculated for that watershed relate to the availability, quality, and quantity of water in your community?

Table 1—Major U.S. Watersheds

1	2	3	4	5	6	7	8	9
Watershed Name	Annual Discharge (km³)	Watershed Area (km²)	Runoff (km³/km²)	Runoff (cm³/cm²)	Annual Precipita-tion (cm)	Evapotranspira-tion + Infiltration (cm)	Percent Runoff	Climate
Where to get data ->	*Gauging Stations*	*Major US Watersheds*	*Calculate*	*Calculate*	*Major US Watersheds*	*Calculated (Precipitation - Runoff)*	*Calculate*	
Mississippi River Basin								humid-continental to humid subtropical
Nelson River Basin								sub-arctic
St. Lawrence River Basin								humid-continental
Colorado River Basin								arid
Columbia River Basin								semi-arid to temperate marine
Rio Grande Basin								semi-arid

Activity 2.6

Do it on the web!

A web-based version of this activity with direct links to the web sites can be found at:

http://saguaro.geo.arizona.edu/ ewr/act26.html

Broken links?

If the web links in this activity do not work, alternate links can be found at:

http://saguaro.geo.arizona.edu/ ewr/act26.html

The local water picture

In this unit, you have looked at long-term weather (i.e. climate) and surface runoff patterns across the United States. You have also examined some of the major weather disasters of the last 20 years and considered the ways in which these disasters affect both local and regional communities. As you will recall, precipitation occurs when uplift and subsequent cooling "wring" moisture out of the atmosphere.

Uplift can be triggered by:
- topography
- weather fronts
- winds (jet stream)
- convection (non-frontal thunderstorms)

To conclude this unit, you will examine the precipitation patterns across the United States in the last 24 hours to determine which of the four factors listed above have influenced recent precipitation events.

Topography

Topography is a constant factor unlike the other three factors listed above. Here, you will consider parts of the country where orographic uplift may play an important role in precipitation.

1. On Map 1, circle areas where precipitation events may be attributed to orographic uplift.

Map 1—Orographic uplift and precipitation

Weather Fronts

In the Explain section, you were introduced to the concept of weather fronts, the boundaries between air masses with different temperature and moisture characteristics. Atmospheric uplift and precipitation are often, but not always, associated with weather fronts, so you will determine the locations of the weather fronts across the country in the last 24 hours, without paying too much attention to the type of front.

▶ Open the following link in your web browser
   http://www.rap.ucar.edu/weather/
▶ Scroll down the page and click the **Weather for today** map.
▶ On the **Analysis and 12-48 hour Prog Charts** page, click the current analysis map (upper left) to open a larger version of the map.

The current positions of fronts are identified in red and blue lines for warm and cold fronts, respectively. On Map 2, note the locations of these fronts.

2. Mark the locations of all weather fronts on Map 2.

Map 2—Weather fronts

▶ When you have finished, click your web browser's **Back** button twice to return to the **Weather Home** page.

Jet Streams

In this activity, you also learned about the significance of jet streams in producing conditions in the upper atmosphere that promote the development of storm systems and in predicting the paths those storms will follow.

▶ On the navigation bar of the **Weather Home** page, click the **Numerical Model** link.
▶ On the **Numerical Model** page, click the **00 hr Forecast** check box and click the **300mb Winds** link.

The map that opens shows the direction and velocity of the jet stream. A scale bar at the bottom of the image shows the wind speeds, with the fastest winds (red and pinks) indicating the location of the jet stream. Remember that there may be more than one jet stream.

3. On Map 3, mark the position(s) of the jet stream(s).

Map 3—Jet streams

▶ When you have finished, click your web browser's **Back** button twice to return to the **Weather Home** page.

Convection (non-frontal thunderstorms)

Another source of atmospheric uplift is convection, the thermal updrafts associated with surface heating and atmospheric instability. In this unit, you learned that thunderstorm activity is typically associated with convection or with frontal uplift. One defining characteristic of thunderstorms is their great height. The cumulonimbus cloud formations associated with thunderstorms often rise to sixty thousand feet or higher in the atmosphere. Because of their height, the tops of thunderstorms are often quite cold relative to the tops of other cloud formations lower in the atmosphere. This temperature difference can be seen in images taken by weather satellites that use color enhanced infrared imagery, in which different colors are used to represent temperature variations in cloud formations.

▶ On the navigation bar of the **Weather Home** page, click the **Satellite** link.

▶ On the satellite image page, click the **Infrared (color)** and **Loop–small** buttons, then click the text **Contiguous U.S.** at the top of the map (see left).

The scale bar at the bottom of the image is in degrees Celsius, with the red colors denoting temperatures above 0° Celsius, and the greens, blues, and purples representing progressively cooler temperatures below 0° Celsius. The strongest convection is indicated by the blues and purples.

4. On Map 4, mark the locations where the strongest convection is occurring *away from the frontal boundaries.*

Map 4—Non-frontal convection

Where to click

Click the words **Contiguous U.S.** at the top of the map. If you click anywhere else on the map, the satellite image will show only a portion of the U.S.

Putting it all together

Now that you have investigated the potential locations that the four factors associated with atmospheric uplift may have influenced precipitation recently, you will examine the actual precipitation events that have occurred over the last 24 hours.

▶ Open the following link in your web browser **http://www.intellicast.com**

▶ Choose **Forecasts** (under **U.S. Weather**), and from the **Historic** menu choose **Daily Precipitation**.

5. On Map 5, identify areas of the U.S. that received precipitation over the last 24 hours. You may want to use different patterns or colored pencils to distinguish the severity of precipitation events.

Map 5—Precipitation events in the past 24 hours

Evaluate, based on the information you obtained, which of these factors appears to be most strongly related to recent precipitation events in the United States. Remember that each of these factors—topography, fronts, the jet streams, and convection—may or may not be responsible for recent precipitation events.

6. Based on the precipitation that has occurred in the U.S. over the last 24 hours, which uplift factors appear to have played an important role in the precipitation? Label them in the appropriate areas on Map 5.

7. Which uplift factors played a relatively minor role in the precipitation that occurred over the last 24 hours?

8. Were there any jet streams over the U.S. in the images you examined? If so, thinking back to the jet stream movie and the four jet stream seasonal positions, is the jet stream where it is supposed to be?

For example...

If you are answering question 9 in January, and you see widespread thunderstorm activity, that is something unexpected. You may need to dig a little deeper to see if other factors might be influencing recent precipitation.

9. Are the factors responsible for recent precipitation events the ones that you would expect for this time of year? Explain.

Unit 3
Using Water Wisely

In this unit, you will ...

- *explore the many ways we use water in daily life;*

- *discover which human activities consume the most water;*

- *investigate regional patterns in water use and the relationship between water use and population size;*

- *identify water use sectors and examine the differences in water use and consumption;*

- *estimate the lifetime of one of the nation's largest aquifers.*

Gene Alexander, USDA NRCS

Center-pivot irrigation system in Colorado.

Activity 3.1   # Water in your world

Water is one of Earth's most plentiful resources, and one of the most widely used. In earlier units, you explored global water reservoirs and the geographic availability of water for human use. In this unit, you will examine the many ways we use water.

Water at home

Your investigation of water's many uses will begin at home. What are some of the ways in which you use water every day? In addition to direct uses of water, consider how water might be used in making or growing the many products you use.

    1. List as many different uses of water in and around your home as you can. (You may attach additional sheets if necessary.)

Water at work and play

Interview five family members, neighbors, or relatives to learn how they use water outside the home, both at work and for recreation. Be prepared to share your lists with your classmates.

    2. List as many different uses of water outside the home as you can.

Categorizing water use

Using the lists you compiled, or sharing your list with a small group or your entire class, develop a list of broad categories that could be used to classify water use. For example, you might categorize the use according to where the water is used, such as "household", or how the water is used, such as "washing." Provide a brief description of each category. Be prepared to share your list of categories with your class.

3. List and describe your major categories of water use. (At least five but not more than ten categories.)

Activity 3.2

Water for many uses

The most accessible reservoirs of fresh water are found on the surface, in lakes and rivers, and below the surface, as ground water. In Unit 1, you learned that these reservoirs are fed by precipitation as water moves through the hydrologic cycle. In Unit 2, you examined the factors that determine where precipitation occurs and how it flows. In this unit, you will examine how fresh water reservoirs are used in the United States. Although water is a renewable resource, it is critical that we do not extract water from reservoirs faster than it can be replenished through the hydrologic cycle. As you explore these data, watch for patterns that illustrate challenges we face in managing our water resources.

Who uses the most water?

▶ Launch ArcView and locate and open the **uswater.apr** project file.
▶ Open the **Water use by state** view.

This view shows the total amount of water—fresh and salt water, surface and ground water—used by each of the 48 contiguous states.

Where are the data for Alaska and Hawaii?

Unfortunately, water use data for Alaska and Hawaii are not available from government agencies in the same format as for the other 48 states and therefore are not included.

1. Does the total water use in the U.S. show any general geographic patterns? Do northern states use more water than southern states? Do eastern states use more than western states? Large states more than small states?

You will examine the data more closely to look for additional patterns.

▶ Activate the **Total water use (mgal/d)** theme.
▶ Click the Open Theme Table button 📊 to open the **Total water use (mgal/d)** theme table.

The **Total water use (mgal/d)** theme table lists water use statistics for each state. Population is given in thousands of people, area in square kilometers, and water use in millions of gallons per day (mgal/d). First, you will look for general patterns in water use among the states.

To activate a theme, click its name in the Table of Contents.

▶ Click the **Total use (mgal/d)** field heading. The field name will be highlighted gray to show that it is selected.
▶ Click the Sort Descending button 📄.

The states are now sorted from highest to lowest total water use.

2. In Table 1 on the next page, list the six states that use the most water.

To select a table field heading, scroll across and click its name. Active fields are highlighted gray.

▶ Select the **Population (thousands)** field heading and sort the states by population, in descending order.

3. In Table 1, list the six states with the largest population.

▶ Select the **Area (sq km)** field heading and click the Sort Descending button 🗐.

4. In Table 1, list the six states with the largest area.

Table 1—Six states with the highest water use, population, and area

States that use the most water	States with the largest population	States with the largest area

5. Using Table 1, determine whether water is better explained by a state's population size or area? Explain.

▶ Select the **Total use (mgal/d)** field heading and click the Sort Ascending button 🗐.

The states are now sorted from lowest to highest total water use.

Table 2—Six states with the lowest water use, population, and area

States that use the least water	Population Rank [Population Rank]	Area Rank [Area Rank]

6. In Table 2, list the six states that use the least water, from highest to lowest water use.

7. Using the information in the theme table, record the population rank and area rank of each state in Table 2.

8. For the states that use the least amount of water, which factor better explains water use—population size or area?

9. What other factors might explain the differences in the amount of water that states use?

▶ Close the **Total water use (mgal/d)** theme table.

Per capita water use

Based on what you have seen, it does not seem appropriate to compare the amount of water a state uses without considering the size of the state or the number of users. To adjust for differences in population size, it is best to compare the *per capita* water use, which is the average amount of water used by each person. The per capita water use is calculated by dividing the amount of water used by the number of users. Per capita use is often expressed in **gpcd**—gallons per capita per day.

▶ Turn off the **Total water use (mgal/d)** theme and turn on and activate the **Per capita total use (gpcd)** theme.

▶ Click the Open Theme Table button 📖 to open the **Per capita total use (gpcd)** theme table.

▶ Select the **Per Capita (gpcd)** field heading and click the Sort Descending button ⬇. Scroll through the table as necessary to complete Table 3.

Per capita?

Per capita is a Latin term that means "for each person."

To turn a theme on or off, click its check box in the Table of Contents.

To activate a theme, click its name in the Table of Contents.

Table 3—Highest and lowest per capita total water use

	State	Per capita total water use rate (gallons/person/day)
Highest		
Your State		
Lowest		

10. In Table 3, record the names and per capita use of the states with the highest and lowest rates of total water use.

11. Why might per capita total water use be a poor choice for comparing the amount of water used by people in different states?

Hint for question 11

Think back to Activity 3.1 where you explored some of the *indirect* ways in which you use water.

▶ Close the **Per capita total use (gpcd)** theme table.

Rather than comparing total water use among states, you can compare the water used just for household or *domestic* purposes. Domestic use includes water that is used in your home.

▶ Turn off the **Per capita total water use (gpcd)** theme and turn on the **Per capita domestic use (gpcd)** theme.

12. Describe any geographic patterns you see in per capita domestic water use.

To select a table field heading, scroll across and click its name. Active fields are highlighted gray.

Self-supplied water

In rural areas, many people have their own water well. States generally do not monitor the water withdrawn from private wells. Therefore, the data for domestic and farm use in these areas is underreported. If a home or farm in a rural area receives water from a public supply (surface water or ground water) that will be reported accurately. In many western states, irrigation water is drawn from public supplies (rivers, etc.), where withdrawals can be measured.

▶ Activate the **Per capita domestic use (gpcd)** theme.
▶ Click the Open Theme Table button 🖩 to open the **Per capita domestic use (gpcd)** theme table.
▶ Select the **Per Capita Use (gpcd)** field heading and click the Sort Descending button 📄.

13. In Table 4, record the names and per capita domestic use for the states with the highest and lowest rates of domestic water use, as well as your own state.

Table 4—Highest and lowest per capita domestic water use rate

	State	Per capita domestic water use rate (gallons/person/day)
Highest		
Your State		
Lowest		

14. What percentage of your state's total water use is domestic use?
% of Total Use = Per capita domestic use (from Question 13) ÷ Per capita total use (from Question 10) × 100

15. What percentage of your state's total water use is *not* for domestic purposes?

16. If this water is not used for domestic purposes, for what kinds of activities is this water used in your state?

To turn a theme on or off, click its check box in the Table of Contents.

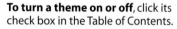
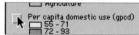

▶ Close the **Per capita domestic use (gpcd)** theme table and turn off the **Per capita domestic use (gpcd)** theme.

Water use by sector

This part of the investigation examines water use in greater detail. The government classifies water use into several major categories, or *sectors*. The major water use sectors are:

- **Commercial**—water used by commercial facilities and institutions including hotels, restaurants, hospitals and schools.
- **Domestic**—water used for household purposes.
- **Industrial**—water used in the production of steel, chemicals, paper, plastics, minerals, petroleum, and other products.
- **Power**—water used to power steam-driven electric generators. It does not include water used for generating hydroelectric power.
- **Mining**—water used for extracting minerals, oil, and natural gas.
- **Agriculture**—water used to raise animals and irrigate crops.

Water for power

Over 99.5% of the water used for generating electrical power comes from surface water—sea water, rivers, streams, and lakes.

A simple way to visualize the amount of water used for different purposes is a pie chart. The pie chart below represents the total amount of water used by each sector in the 48 contiguous states. The entire circle represents the total amount of water used for all purposes, and the slices represent the percentage of the total water used by each of the six water use sectors.

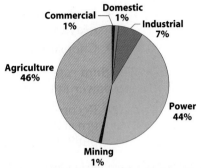

17. According to the chart above, which two sectors use most of the water in the U.S.? Combined, how much of the total water do they use?

East versus West

To activate a theme, click its name in the Table of Contents.

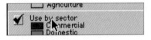

▸ Turn on and activate the **Use by sector (mgal/d)** theme.

This theme shows a water use pie chart for each state. The amount of water used by each state is represented by the size of its pie chart.

18. In the western states, what sector makes up the largest piece of each pie?

19. In the eastern states, what sector makes up the largest piece of each pie?

Next you will compare the amount of water used by the power and agriculture sectors in the eastern and western United States.

Why is nothing highlighted when I make the selection?

In the **Use by sector (mgal/d)** theme, the states are transparent to allow you to see other themes behind the pie charts. Therefore, there is nothing to highlight. You will know if you selected correctly in the next step.

▸ Use the Select Features tool 🔲 to select the 17 western states, as shown below. Do not touch any states east of Texas.

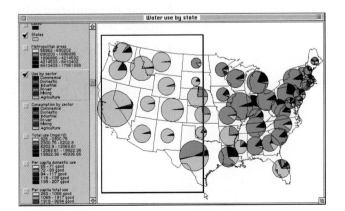

▸ Click the Open Theme Table button 🖩 to open the **Use by sector (mgal/d)** theme table. The status bar should read "**17** of **48**," indicating that you have 17 of the 48 states selected. If not, close the theme table and select the western states again.

▸ Click the **Power** field heading. The field name will be highlighted gray to show that it is selected.

▸ Choose **Field ▸ Statistics...**

The **Sum** reported in the statistics window is the water used for generating power in the western states in millions of gallons per day.

20. Record the amount of water used for power production in the western states in Table 5, round to the nearest whole number. (Be sure to put it in the column labeled Water Use, not Water Consumption).

Table 5—Comparing water use and consumption in western and eastern states

States	Population (thousands)	Water Use (mgal/d)		Water Consumption (mgal/d)		% Consumption	
		Agriculture	Power	Agriculture	Power	Agriculture	Power
Western							
Eastern							

▸ Repeat this procedure to find the amount of water used for agriculture [**Agriculture**] and the total population size [**Population (thousands)**]] in the western states.

21. Record the amount of water used for agriculture and the population of the western states in Table 5. Round all values to the nearest whole number.

▸ To select the 31 eastern states, simply click the Toggle Selection button ♻. The status bar should read "**31** of **48.**"

▸ Repeat the procedures above to find the population and the amount of water used for power and agriculture in the eastern states.

22. Record the total population size and the amount of water used for power and agriculture in the eastern states in Table 5, round all values to the nearest whole number.

▸ Close the statistics window and the **Use by sector (mgal/d)** theme table.

▸ Choose **Theme ▸ Clear Selected Features**.

23. What is the relationship between population size and amount of water used by the agriculture and power sectors in the western states? What about in the eastern states?

Use versus consumption

Not all water that is utilized by various water use sectors is completely *consumed*, or "used up". Often, water used for one purpose can be used again. For example, water used by power plants for cooling can be reused or returned to fresh water reservoirs. Other types of wastewater can also be treated and reused. Water that is returned for reuse is called *return flow*. By subtracting the return flow from the total water used, you can find out how much water has actually been consumed.

To turn a theme on or off, click its check box in the Table of Contents.

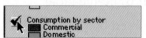

▶ Turn on the **Consumption by sector (mgal/d)** theme.

This theme also uses pie charts to show the amount of water consumed by the six water use sectors in each state. The diameter of the pie chart represents the total amount of water consumed by that state.

▶ Leaving the **Use by sector (mgal/d)** theme on, turn the **Consumption by sector (mgal/d)** theme on and off, comparing the two sets of pie charts.

24. Are the pie charts for water consumption larger or smaller than the pie charts for water use?

25. In which water use sector do you see the largest difference when you compare water use to water consumption? Explain.

▶ Turn off the **Use by sector (mgal/d)** theme.

To activate a theme, click its name in the Table of Contents.

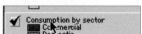

▶ Activate the **Consumption by sector (mgal/d)** theme.

▶ Use the Select Features tool 🔲 to select the seventeen western states, as you did on page 77. Be sure your selection rectangle does not touch any states east of Texas.

▶ Click the Open Theme Table button 🖻 to open the **Consumption by sector (mgal/d)** theme table.

The status bar should again read "**17** of **48**," indicating that you have 17 of the 48 states selected. If not, close the theme table and select the western states again.

▶ Click the **Power** field heading. The field name will be highlighted gray to show that it is selected.
▶ Choose **Field ▶ Statistics...**

The **Sum** is the total amount of water, in millions of gallons per day, *consumed* in the process of producing power in the western states.

26. Record the total amount of water consumed for power production in the western states in Table 5. Round to the nearest whole number.

▶ Close the statistics window.

▶ Repeat this procedure to find the total amount of water consumed by agriculture (**Agriculture**) in the western states.

27. Record the total amount of water consumed by agriculture in the western states in Table 5.

▶ Click the Toggle Selection button ⟳ to select the 31 eastern states. The status bar should read "**31** of **48**."

▶ Repeat the procedures above to determine the amount of water consumed by the power and agriculture sectors in the eastern states. Round all values to the nearest whole number.

28. Record the amount of water consumed by the power and agriculture sectors in the eastern states in Table 5.

▶ Close the **Consumption by sector (mgal/d)** theme table.

29. Calculate the percentages of water consumed by the agriculture and power sectors in both the eastern states and the western states and enter them in Table 5. Round to the nearest 1%. Use the formula *% Consumed = water consumed / water used × 100.*

30. Examine Table 5. Which of the two water use sectors consumes the highest percentage of the water it uses? Explain why you think this occurs.

Activity 3.3

Water at work

There is no overall shortage of water on Earth. The continuous movement of water through the hydrologic cycle provides a renewable resource. However, water of sufficient quality and quantity may not be available when and where it is needed. Much of the water we consume is ground water drawn from underground aquifers that takes tens of thousands of years to accumulate. Using ground water faster than it is renewed by the hydrologic cycle creates an imbalance. For these reasons, water is considered a finite, or limited, resource, and balancing the many demands for water with the available supply is an ongoing challenge.

How is water used?

To make good decisions about water usage, you have to understand the major usage categories—domestic, commercial, industrial, mining, power generation, and agricultural. You will take a closer look at how water is used in each of these sectors. As you explore these sectors in detail, consider the questions "How is water being used?" and "How *should* water be used?"

Domestic

Domestic use refers to the water we use in our homes, for things like cooking, drinking, bathing, washing clothes, flushing the toilet, brushing your teeth, watering the lawn, and so on.

Water is most often supplied to our homes by a city or county water department or by a private company. This supply is called the public supply. In 1995, about 85% of the water used for domestic and commercial purposes came from the public supply. The remainder was self-supplied water from private wells, particularly in rural areas.

Commercial

Commercial water use includes the same activities as domestic use, but applied to commercial businesses. The water required to prepare a meal at a restaurant—to wash produce, mix the syrup for a soda, clean the floors—is considered commercial use. Water used by the military, schools, prisons,

L.P. Kendall, The SAGUARO Project

Figure 1. The water used in this self-service car wash is included in the commercial sector.

businesses, and hotels is also considered commercial use. Commercial water usually comes from the public supply, although some commercial users have their own wells. Golf courses, for example, are often irrigated with water from private wells.

Industrial

In industry, water is used to manufacture virtually every product you own or use. In addition to water used as an ingredient in the manufactured product, water is also used for washing and cooling machinery used during the manufacturing process. Industrial water quality requirements are often less strict than in the domestic and commercial sectors.

Surface water accounts for more than 75% of the water used for industrial purposes. This contrasts sharply with the domestic sector, where ground water is strongly preferred due to consumers' demand for "pure" water. Industry is sometimes able to make use of salt water, which is almost never used in the domestic market.

Industrial water users have an important advantage over other users: water used for cooling and cleaning can often be reused. High quality filters remove impurities, permitting the reuse of cleaning water. In cooling systems, water needs only to cool down before it can be used again. Overall, industry consumes only 15% of the water it uses—the other 85% is returned for reuse.

Mining

In mining, water is used during the extraction of minerals like copper and silver, and fossil fuels including coal, oil, and natural gas. Water used in stone quarries and ore mills, and to control dust in and around mines is included in mining water use. Water used in processing minerals and refining fossil fuels is generally reported as industrial use.

Generating power

Electrical power generators require water for cooling, in much the same way that other industries use water for cooling. Additionally, power is generated by using fuel to heat water to steam, which turns massive turbines that generate electricity. After cooling, this water is either reused or returned to the environment. Nearly all of the water used for power generation is surface water drawn from lakes and rivers.

Water is also used to produce electricity in hydro-electric power plants. Rivers provide the moving water necessary to drive generator turbines. To provide sufficient water pressure, dams are often built, creating artificial lakes. Using the water stored in these lakes, we can generate a steady supply of power despite seasonal changes in flow, and produce additional power to meet increased demands during the hot summer months. Water in these lakes and reservoirs is also frequently used by nearby farms and cities.

Agriculture

Water is used in a variety of ways in agriculture, including irrigating crops, raising livestock, and mixing chemicals to apply to crops. This section will focus on the two major uses, crop irrigation and livestock support.

Irrigation

Irrigation uses more water in the United States and around the world than any other activity. Without the ability to supply water to millions of acres of crops, modern agriculture would grind to a halt. The majority of the croplands in the U.S. are located in areas that do not receive reliable precipitation, and must either import surface water from great distances or pump ground water to irrigate crops. Approximately 37% of the water used for irrigation in the U.S. is ground water, and 50% of this water is used by just five states: California, Idaho, Colorado, Texas, and Montana.

Nearly half of the water used for irrigation is lost through leaking pipes and evaporation. Evaporation is the largest source of water loss. Most irrigation involves spraying water into the air over crops. The small water droplets have a large surface area to volume ratio, resulting in higher evaporation rates compared to large bodies of standing water.

Livestock support

In the care and feeding of livestock, drinking water is one obvious use of water. The quantity of water used by farm animals varies greatly depending on the size of the animal. Chickens require very little drinking water (less than a cup per day), while cattle drink much more (several gallons per day, though milk-producing cows require even more drinking water than beef cattle).

Water is also used to process food produced from the animals (meat, eggs, milk, and cheese) and manage animal waste. Management of animal waste creates many environmental problems. This wastewater is typically stored in open-air lagoons. If water from these lagoons enters our fresh water supplies, it poses a serious health risk. During Hurricane Floyd in 1999, many of the fresh water reservoirs in North Carolina were contaminated when lagoons filled with hog waste were flooded by the heavy rainfall and the waste water was spread throughout the state.

Water sustainability

The overall amount of water used in the United States has remained constant or even decreased slightly since about 1985. In the domestic use sector, per capita water use has decreased due to conservation measures, however total use has increased by 14% due to population growth. Our growing population will eventually overtake our ability to further reduce water use through conservation, and we will be faced with increasingly difficult challenges and decisions. Because water supply and demand are not evenly distributed across the country, some areas will have to deal with more complicated water issues than other areas.

Management of water resources involves both maintaining water quality and determining how water should be used. Since clean, safe drinking water is essential, water quality is of great concern. Cleaning up polluted water supplies is both difficult and expensive. Where contamination has already occurred, cleaning up the mess is our only option. However, it is equally important that we take preventative measures to protect our water supply from further contamination by harmful pollutants.

Issues of water management and sustainability are not unique to the United States. All nations face challenges in providing sufficient amounts of clean, safe water to the public. Global population growth and its relationship to the water supply involve complex, interconnected issues that will figure increasingly in world affairs.

Three Rs of water conservation

There are steps we can take to conserve water. These are sometimes referred to as the three Rs of conservation—Reduce, Reuse, and Recycle. Here are some of the ways in which people save water in each of the major use sectors.

The decline in per capita domestic water use in the United States since 1985 has been a result of conservation efforts such as low-flow toilets and showers, improved home irrigation systems and habits, and effective water conservation education programs. According to studies by the Environmental Protection Agency, the average family of four now uses 20,000 fewer gallons a year than they did even a few years ago. That is enough water to fill a small backyard swimming pool!

The commercial, industrial, and mining use sectors have reduced water use significantly in recent years. New laws and financial incentives have encouraged businesses to develop more efficient processes and practices that use less water, and use more recycled water. Car washes and factories collect and filter waste water and reuse it many times before it is discharged into the sewer system.

Although electric power generating plants use tremendous quantities of water for cooling, virtually all of the water withdrawn is returned or reused many times. Most power plants are built near surface water sources, and are capable of using "low-quality" water that is not suitable for domestic or agricultural use. In fact, almost one-third of the water used in generating electricity is salt water.

Recall that roughly half of the water used to irrigate crops is wasted. The waste can be reduced by using alternative irrigation methods (Figure 2). Drip-irrigation systems use one-third as much water as traditional spray-irrigation systems, because they irrigate plants directly rather than by

Figure 2. Irrigation methods

Flood or furrow irrigation—Where water is relatively cheap, flood irrigation is the simplest and least expensive irrigation method. It requires very flat fields to avoid ponding. Flooding is a good way to flush salts out of the soil, but evaporation losses are very high.

Spray—More expensive than flood irrigation, spraying is used on crops where flooding is impractical. Evaporation losses are high, and applying water to leaves encourages the growth of plant diseases.

Drip or microirrigation—The most expensive, but also the most water-efficient irrigation technique. Requires high-quality water (low in salts) to prevent clogging and the buildup of salinity in soil around plants. Drip irrigation is most often used in orchards and vineyards where watering is concentrated on individual trees and the ground is not plowed each year.

spraying water into the air. Drip irrigation is not practical or appropriate for many crops, but it is very effective in orchards and vineyards that are not plowed and replanted each year.

Another step suggested by some conservationists is to reduce the quantity of meat in our diets, thus reducing the quantity of water used in the production of livestock. Water is used to grow feed, clean up waste, and process livestock into food. It has been estimated that it takes as much as 12,000 gallons of water to produce a pound of beef, compared to about 110 gallons to produce a pound of wheat. Such a suggestion would certainly not be supported by the livestock industry or people who enjoy meat products. Clearly, it is a complex issue.

Questions

1. Describe a use of water in the commercial sector.

2. Describe three ways in which business and industry help to conserve fresh water.

3. Which source supplies the most water to industries—ground water or surface water? Why?

4. Which types of crops are best suited to drip irrigation? Explain.

5. What percentage of the water used for irrigation is lost? Why?

6. Why has per capita domestic water use in the United States decreased, even though the population has increased?

Activity 3.4

Food exports

In addition to producing food for our own consumption, U.S. farmers and ranchers export many agricultural products to other countries. In 2000, these products accounted for $52 billion, or 6.7% of our total exports.

Reading the precipitation legend

To see the **Precipitation (cm/yr)** theme legend, scroll to the bottom of the Table of Contents.

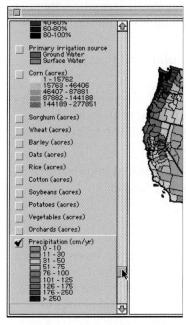

To turn a theme on or off, click its check box in the Table of Contents.

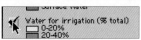

Missing county data

Some counties either did not collect or supply water data or were allowed to withhold their data to protect the economic interests of their farmers.

Feeding a nation

In an earlier activity, you learned about six major water use sectors—domestic, commercial, industrial, mining, power, and agriculture. Of these, agriculture consumes by far the most water. In this activity, you will examine patterns in agricultural water use and discuss the implications of those patterns.

Precipitation and irrigation

▶ Launch ArcView and locate and open the **uswater.apr** project file.

▶ Open the **Water for agriculture** view.

This view shows the average annual precipitation for each county of the contiguous United States, in centimeters per year. The climate of a region is determined by the amount of precipitation it receives annually.

Climate Type	Precipitation
Arid	30 cm or less
Semi-Arid	31-50 cm
Sub-Humid	51-75 cm
Humid	76-200 cm
Tropical Humid	more than 200 cm

▶ Scroll down the Table of contents to read the precipitation legend.

1. Based on precipitation, how would you classify the climate of the western U.S.? How would you classify the climate of the eastern U.S.?

The climate difference between the eastern and western half of the country has shaped the development of agriculture in the U.S. The East has adequate rainfall to support all but the most water-thirsty crops. In the West, precipitation is unreliable and sufficient amounts of surface water are not available. Therefore, large-scale agriculture depends on ground water to irrigate crops.

▶ Turn on the **Water for irrigation (% total)** theme.

This theme shows the percentage of each county's total water consumption that is used for irrigation. Notice that some counties use nearly all of their water for irrigation.

▶ Turn the **Water for irrigation (% total)** theme off and on several times, and look for patterns related to precipitation.

You can see that areas with low precipitation depend heavily on irrigation. However, if you look closely you will also notice areas that are extensively irrigated, even though those areas receive significant rainfall (over 125 cm) each year.

2. On the map below, indicate three regions that use large amounts of water for irrigation, yet also receive more than 125 cm of precipitation per year.

This view includes themes that show where major crops are grown. Next, you will compare individual crop themes to the irrigation theme to find out which crops are grown in the areas where the natural precipitation is supplemented by irrigation.

▶ Turn off the **Water for irrigation (% total)** and **Precipitation (cm/yr)** themes.

To turn a theme on or off, click its check box in the Table of Contents.

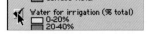

▶ Turn each individual crop theme on and off, beginning with **Sorghum (acres)** and ending with **Orchards (acres)**. Examine the areas you identified on the map above to see if the crop is grown in that area.

3. On the map above, label each area you identified in question 2 with the names of the crops grown in that region.

Agricultural water sources

Subsidized irrigation water

The Bureau of Reclamation provides about one quarter of the irrigation water in the West at wholesale prices. Water quantities and prices are set by long-term contracts, and are highly subsidized by taxpayers. Farmers are charged an annual fee according to the acreage of crops served, rather than the actual amount of water used. Subsidies average $54 per acre, but are as high as $150 per acre.

In some areas, there is enough water in nearby lakes and rivers to meet irrigation needs. The cost to pump and deliver this "on-farm" surface water ranges from free to about $15 per acre per year. In the arid West, the Bureau of Reclamation has built dams, reservoirs, and canals to capture and deliver water to farms. This "off-farm" surface water costs $10 to $90 per acre, but is heavily subsidized by the government. In areas with expensive or unreliable surface water supplies, farmers may pump ground water. Due to the high cost of pumping, ground water costs from $11 to $105 per acre. Many farmers use a combination of ground and surface water to meet their irrigation needs.

▶ Turn off everything except the **States** theme.
▶ Turn on the **Primary irrigation source** theme.

This theme represents the primary source of water used for irrigation. Counties colored brown use ground water and counties colored blue use surface water as the primary source for irrigation.

4. On the map below, mark at least three large regions *east* of the Rocky Mountains that depend on ground water for irrigation.

What is an aquifer?

An aquifer is a water-bearing rock formation.

To turn a theme on or off, click its check box in the Table of Contents.

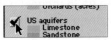

To activate a theme, click its name in the Table of Contents.

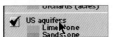

▶ Turn on and activate the **US aquifers** theme.

▶ Use the Identify tool ⓘ to determine the names of the aquifers that provide ground water for the regions you marked on the map.

5. Label each of the regions you marked on the map above with the name of the major aquifer that provides irrigation water.

▶ Turn off the **US aquifers** and **Primary irrigation source** themes.

The High Plains Aquifer

The largest of these aquifers, the High Plains Aquifer—also called the Ogallala Aquifer—covers some 450,000 square kilometers (174,000 square miles) in Texas, Oklahoma, Nebraska, New Mexico, Colorado, Kansas, and South Dakota.

▶ Turn on the **High Plains Aquifer** and the **Crops by county (% acreage)** themes.

The **Crops by county (% acreage)** theme displays small pie charts showing the percentage of farm acreage used for nine major crops. You can get a good idea of the major crops grown in different regions by looking at the color patterns formed by the pie charts.

▶ Use the Zoom In tool 🔍 to zoom in on the High Plains Aquifer.

6. What are four major crops irrigated by the High Plains Aquifer?

Agricultural scientists have developed the crop coefficient (Kc) as a measure of the water needs of different crops. This measurement compares the amount of water needed by each crop to a standard crop such as grass or alfalfa under the same conditions. Farmers use crop coefficients, along with average temperature, latitude, and other climate factors to determine their irrigation needs.

Crop coefficients (Kc)

In this chart, bermuda grass is the standard crop, with a Kc value of 1.

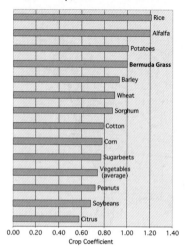

7. According to the crop coefficient chart, how do the water needs of the crops served by the High Plains Aquifer compare with the water needs of other crops?

▶ Turn off the **Crops by county (% acreage)** theme.

▶ Choose **View ▶ Full Extent** to view the entire map again.

Squeezing the sponge

The High Plains Aquifer is composed of layers of sand, clay, and gravel eroded from the Rocky Mountains. Water is stored in the sand and gravel layers. The thickness of the water-bearing sediments of the aquifer ranges from less than a meter to several hundred meters.

When farmers first tapped this water source in the late 1930s, they thought it would provide an unlimited supply of water. However, the overlying rock is fairly impermeable, it allows very little water to infiltrate the ground and reach the aquifer. Consequently, the aquifer's recharge rate—the rate at which water is replenished from the surface—is very low. In fact, the recharge rate averages only about 0.9 cm (0.35 inches) per year across the aquifer. Next, you will calculate how long the aquifer can support agriculture at the current rate of withdrawal.

▶ Turn on and activate the **US aquifers** theme.

To turn a theme on or off, click its check box in the Table of Contents.

To activate a theme, click its name in the Table of Contents.

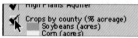

▶ Click on the High Plains Aquifer using the Select Feature tool 🔲 to select it. The aquifer will be highlighted yellow.

▶ Turn on and activate the **Primary irrigation source** theme.

▶ Choose **Theme ▶ Select By Theme...** and choose from the pop-up menus to select features of the active themes that **Have their Center In** the selected features of the **US aquifers** theme. Click the **New Set** button.

```
╔═══════════════ Select By Theme ═══════════════╗
║                                                ║
║  Select features of active themes that         ║
║                                                ║
║  ┌──────────────────────────┬─▼─┐  ╭────────────╮ ║
║  │ Have their Center In      │   │  │  New Set   │ ║
║  └──────────────────────────┴───┘  ╰────────────╯ ║
║  the selected features of          ╭────────────╮ ║
║                                    │  Add to Set │ ║
║  ┌──────────────────────────┬─▼─┐  ╰────────────╯ ║
║  │ US aquifers               │   │  ╭────────────╮ ║
║  └──────────────────────────┴───┘  │Select from Set│║
║                                    ╰────────────╯ ║
║                                                ║
║                                    ╭────────────╮ ║
║                                    │  Cancel     │ ║
║                                    ╰────────────╯ ║
╚════════════════════════════════════════════════╝
```

This will select the counties that lie mostly within the boundaries of the High Plains Aquifer. Next, you will determine the amount of ground water used for irrigation in these counties.

▶ Click the Open Theme Table button 🖼 to open the **Primary irrigation source** theme table.

▶ Click the **Ground (mgal/d)** field heading. The field name will be highlighted gray to show that it is selected.

▶ Choose **Field ▶ Statistics....**

To select a table field, scroll across and click on the field name. Active field names are highlighted gray.

The **Sum** is the total daily ground water withdrawals from the counties served by the aquifer, in millions of gallons per day (mgal/d).

8. How many million gallons of ground water are withdrawn from the aquifer each day?

Converting from gallons to cubic meters

1 cubic meter = 264 gallons

so,

millions of gallons ÷ 264 g/m³ = millions of cubic meters.

9. Convert the withdrawal rate to cubic meters per year.
   a. Multiply by 1,000,000 to convert from millions of gallons per day to gallons per day. (Hint—just add six zeroes.)

Rounding

If you need a refresher on rounding, see page *v*.

   b. Divide by 264 gal/m³ to convert from gallons per day to cubic meters per day. Round the result to the nearest 100,000 m³/d.

   c. Multiply by 365 days/year to find the annual withdrawal from the aquifer in cubic meters per year. Round to the nearest 100,000 m³/yr.

The total surface area of the aquifer is 450,000,000,000 m², and the recharge rate is 0.009 m/yr.

10. Calculate the total annual recharge by multiplying the area of the aquifer by the recharge rate.

11. What is the difference between the annual rate of withdrawal (question 9c) and the rate of recharge of the High Plains Aquifer (question 10)? Is this an annual gain or loss of water for the aquifer?

The total volume of available water in the High Plains Aquifer is estimated at 4.03 trillion m³. (4,030,000,000,000 m³)

12. For how many more years will the High Plains Aquifer last at the current rates of recharge and withdrawal? (Hint: divide the total volume of the aquifer by the annual loss calculated in question 11.)

13. Do you think this estimate is a realistic one? What factors could cause the depletion of the High Plains Aquifer to occur more rapidly than your estimate?

14. Based on this information, what do you think could realistically be done to protect this immense but finite resource? Propose a series of actions and a time line for implementing those actions. For example, what could you change about the type and acreage of the crops grown there, or the irrigation practices?

Activity 3.5

Meeting the challenge

One of the goals of this unit is to examine patterns of water use and our ability to sustain that use. Here are three primary issues in water management.

- **Quantity** - Is there enough fresh, clean water to meet the various needs of the population?
- **Proximity** - Are the sources of water close to where it is used, or far away?
- **Quality** - Is the water clean and pure enough for its intended use?

Discuss these water challenges as a class or in small groups and list any additional challenges that you identify. You may wish to re-examine the water use data using ArcView to explore these issues in more detail.

1. List water challenges that your state, county, or city are facing, or will probably face in the near future.

2. How does each of these issues affect your state? Give specific examples from the water use data you have explored in the activity, current events, or personal knowledge and experience.

3. How do these issues affect you personally? How might they change your lifestyle or behavior?

4. If you were in charge of water policy for your state, what measures would you recommend, in the face of these challenges, to ensure an adequate water supply for the next 50 years?

Unit 4
Water for a Desert City

In this unit, you will...

- *identify water-related challenges faced by desert and western cities,*

- *examine the importance of maintaining a water balance,*

- *determine the economic and environmental consequences of excessive ground water pumping,*

- *investigate water use patterns and the importance of water conservation, and*

- *develop a plan to help Tucson, Arizona meet its future water needs.*

Downtown Tucson, Arizona and the Catalina Mountains.

L.P. Kendall, The SAGUARO Project

Activity 4.1

Living in a desert

Desert regions are characterized by low annual precipitation and a high evaporation rate. These regions, covering approximately 20% of the land on Earth, may be large expanses of shifting sand with few plants or rugged mountainous regions colonized by many plant species. Although many people think deserts are uninhabitable, global deserts and their margins contain many cities with thriving populations.

In the United States, the four large deserts of the Southwest—the Chihuahuan, Great Basin, Mojave, and Sonoran deserts—contain many cities with significant populations (Figure 1). One of these cities is Tucson, the second largest metropolitan area in Arizona, with a population of over 800,000. This region was originally occupied by Native Americans and later by Spanish settlers who established missions and military posts beginning in 1775. After Arizona became part of the United States in 1912, the population of Tucson grew as people realized the potential for mining and agriculture. The continued population growth of the Tucson area has had important consequences for the environment. The demand for more water is met by drilling new wells and importing surface water from distant sources.

The four great deserts of the Southwest

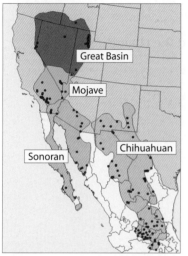

Figure 1. Red dots represent cities in the four major deserts of the southwestern U.S. and Mexico.

1. What water reservoir is tapped by pumping water from wells?

2. Speculate on the environmental consequences of removing water from this reservoir faster than it is replenished by the hydrologic cycle.

3. Do you think that issues of water supply and demand are unique to desert cities and towns? Explain how the problems of obtaining adequate water for desert inhabitants might apply to your city or town.

The Santa Cruz River was a critical water source for early Tucson inhabitants. Archaeologists have found evidence of agricultural settlements along the Santa Cruz dating back to approximately 1000 B.C. Compare the two photographs on page 97 taken of the west branch of the Santa Cruz River. Both photographs were taken from the same location, 96 years apart.

4. Describe the major differences in the environment between the two photographs.

5. What do you think might have caused the changes?

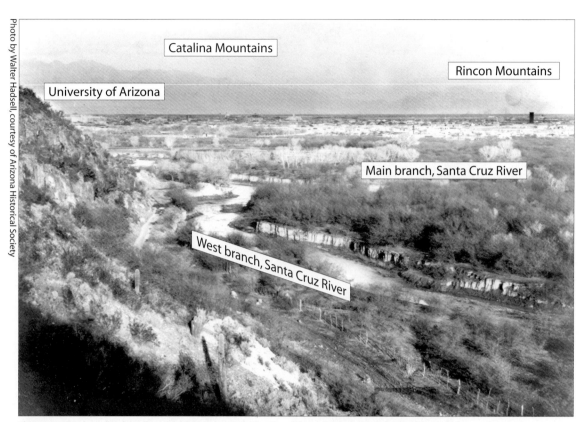

Photo by Walter Hadsell, courtesy of Arizona Historical Society

Figure 2. West branch of the Santa Cruz River in 1904. The city of Tucson, Arizona is in the background.

Dominic Oldershaw, USGS

Figure 3. View from the same location in 2000. The west branch of the Santa Cruz River has been filled in, but the main channel is still visible, above center.

Activity 4.2

Water in the balance

Maintaining a balance between the water we use and the water supply that is replenished by precipitation is the key to ensuring that cities and agriculture continue to thrive. This is a particularly demanding task in desert regions. The deserts of the American Southwest contain bustling cities and an extensive network of farms and ranches that depend on water for survival.

Tucson, Arizona is a desert city that relies on ground water withdrawn from the underlying aquifer to meet almost all its water needs. Therefore, to maintain a water balance, we must understand when, where, and how the aquifer is recharged or replenished. In this activity, you will examine the factors critical to recharging the aquifer.

What is an aquifer?

An **aquifer** is a water-bearing rock formation.

Precipitation patterns in the Tucson Basin

Precipitation events provide most of the water that recharges aquifers. Therefore, examining precipitation patterns is the first step to understanding aquifer recharge.

> ▸ Launch ArcView and locate and open the **tucson_water.apr** project file.

> ▸ Open the **Precipitation Patterns** view.

This view shows a shaded relief image of the Tucson area as well as the city streets.

> ▸ Turn on and activate the **Precip Zones (cm/yr)** theme.

30-year data

The data in the **Precip Zones (cm/yr)** theme are actually averaged over a 30-year period from 1971 to 2000.

The colors that are displayed in this theme correspond to the amount of annual precipitation received by different areas. The heavy dashed line marks the approximate boundary of the Tucson Basin, including the slopes of the mountains that drain into the basin.

> ▸ Using the Identify tool [**i**], examine the amount of precipitation received within each of these zones.

> 1. Is the amount of annual precipitation uniform across the Tucson Basin? Explain your observations.

> ▸ Turn on and activate the **Weather Stations** theme.

This theme shows the location of weather stations within and outside the Tucson Basin.

> ▸ Using the Identify tool [**i**], determine which weather stations inside the Tucson Basin receive the highest and the lowest amounts of annual precipitation.

2. In Table 1, list the weather stations that receive the highest and lowest amount of annual precipitation in the Tucson Basin. Include the amount of annual precipitation they receive and the month they receive the most precipitation.

Table 1—Highest and lowest annual precipitation in the Tucson Basin

	Name of station	Annual precipitation (cm/yr)	Month of highest precipitation
Station with highest precipitation			
Station with lowest precipitation			

▶ Close all Identify Results windows and turn off the **Precip Zones (cm/yr)** and **Weather Stations** themes.

Aquifer recharge

Precipitation is continually replacing, or *recharging* ground water that we remove from the aquifer. Of course, not all precipitation ends up back in the aquifer. As you learned in previous activities, precipitation is diverted through different parts of the hydrologic cycle when it reaches land.

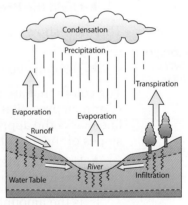

Figure 1. Land processes of the hydrologic cycle.

Watch a flash flood!

To view a movie of a flood caused by summer thunderstorm precipitation runoff, click the Media Viewer button 🖳 and choose **Flood Movie**.

3. Examine the hydrologic cycle diagram (Figure 1) to determine what can happen to precipitation once it reaches the ground. Which of these processes recharges the aquifer?

▶ Turn off the **Streets** theme.

Natural recharge

Precipitation infiltrates both the permeable basin soils and, to a lesser extent, the joints, fractures, and faults in the bedrock of the surrounding mountain ranges. If precipitation falls in a region where the ground is saturated or not permeable, that water becomes runoff and joins streams and rivers. In the Tucson Basin, streams and rivers are typically dry because the water table lies below the river bottoms. After precipitation events, these water channels fill with runoff and merge together as they flow downstream and eventually drain into the Santa Cruz River. The amount of water flowing in rivers and streams is called **discharge**.

▶ Turn on the **Streams** theme.

This theme illustrates the network of streams and rivers that flow through the Tucson Basin. The "V" shape that is formed where two streams merge points downstream, as shown at left.

4. In which direction does water flow out of the Tucson Basin? (On the map, north is "up." Turn on the **Precip Zones (cm/yr)** theme if you need to see the basin boundary.

▶ Turn on and activate the **Gauging Stations** theme.

This theme shows the location of several gauging stations on Tucson area streams. Gauging stations measure the rate of discharge or flow of the water in rivers and streams. You will use the information provided by gauging stations to determine if discharge increases or decreases as the water flows out of the Tucson Basin.

▶ Use the Identify Tool 🛈 to determine the name of each gauging station.

5. Label the gauging stations on the stream diagram below.

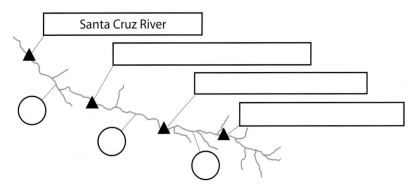

Santa Cruz River

6. In the circles, indicate the number of streams that enter the Rillito Creek between gauging stations.

7. Do you expect the discharge to increase or decrease as the water flows out of the Tucson Basin? Explain your answer.

Permeable surfaces

A permeable surface permits water to pass through it.

Stream flow

The "V" shape that forms where streams merge points downstream.

Gauging station

One type of gauging station is called a **stilling well**, illustrated below.

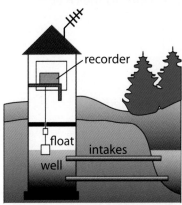

Stream water enters the well through intake pipes. The water level in the well is the same as the water level in the stream. As the water level rises or falls, the float in the well also rises or falls. A cable attached to the float drives a device that records the water level or transmits it to a satellite for recording at a central location. (Adapted from USGS)

▶ Click the Media Viewer button 🔲 and choose the **Discharge Movie** from the media list.

This movie shows the amount of discharge flowing through the four gauging stations on the Rillito Creek and Santa Cruz River as they flow out of the Tucson Basin.

▶ Watch the movie several times to determine whether discharge increases or decreases between each of the stations.

8. How does the discharge change between the following gauging stations? (Does it increase, decrease, or remain constant?)
   a. Tanque Verde Creek and Rillito at Dodge Blvd. –

   b. Rillito Creek at Dodge Blvd. and La Cholla Blvd. –

9. How does this compare to your prediction in Question 7? What process accounts for this observation? Hint: think about what can happen to the water in the hydrologic cycle.

Spiky data

The data in the **Discharge Movie** appear more "spiky" than the precipitation data you looked at earlier, because they are daily measurements rather than monthly averages. Furthermore, the gauging station data are for a single year (1998) rather than a 30-year average.

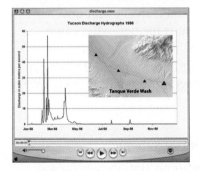

You may have noticed that there is discharge or water flowing year round at the Santa Cruz River gauging station. As noted in the movie, this base flow is a result of the addition of treated waste water from the sewer system.

10. What is the base flow, in m³/sec, at the Santa Cruz River station?

11. What might be the benefit of adding this water to the river system?

▶ Turn off the **Streams** and **Gauging Stations** themes.

Estimating annual recharge

Next you will estimate the amount of water recharged into the Tucson Basin aquifer annually.

▶ Turn on and activate the **Recharge Regions** theme.

The **Recharge Regions** theme shows the three main recharge regions of the Tucson Basin. Only the portion of the mountain slopes that drain into Tucson Basin are included.

▶ Using the Identify tool 🔲, click in each recharge region to gather information to record in Table 2.

12. Record the area and average annual precipitation for each recharge region in Table 2 below.

Table 2—Calculating recharge in the Tucson Basin

Recharge Area Name	Average Precipitation (m) *click in region using the Identify tool*	Area (m²) *click in region using the Identify tool*	Precipitation Volume (m³) *= Area x Average Precipitation*	Recharged Volume (m³) *= Precipitation Volume x 0.05*
Central Basin				
Tucson Mountains				
Catalina–Rincon Mountains				
Totals *(add values in each column)*				

13. Calculate the volume of precipitation that falls in each region by multiplying the Average Precipitation by the Area. Record your results in Table 2.

Research has shown that only about 5% of the precipitation that falls in the Tucson Basin recharges the aquifer.

14. Calculate the volume of water recharged in each region of the Tucson Basin by multiplying the Precipitation Volume by 0.05 (5%) and record your results in Table 2.

15. Calculate the total Area, Precipitation Volume, and Recharge Volume by adding the values in each column of Table 2.

16. Based on what you know about the hydrologic cycle and the climate in Tucson, why do you think the recharge rate (5%) is so low in the Tucson Basin?

The Colorado River— a desert lifeline

Originating in the Rocky Mountains, the Colorado River is a lifeline for the southwestern United States. Most of the territory the river crosses on its way to the Gulf of California is desert. Along the way, water is withdrawn to meet the needs of 20 million people and to irrigate millions of acres of farmland.

What's an acre-foot?

An **acre-foot** is the amount of water needed to cover one acre of land to a depth of 1 foot. One acre foot equals about 325,851 gallons, the amount used by an average family of four in one year.

Artificial recharge

You have explored natural, precipitation-driven recharge of the aquifer and you know that the aquifer is artificially recharged using treated waste water. Surface water imported from other areas is also used to artificially recharge the aquifer. For example, the Central Arizona Project (CAP), completed in 1993 at a total cost of $4 billion, delivers approximately 1.9 billion m³ (1.5 million acre-feet) of Colorado River water to cities in central and southern Arizona each year. The CAP canal carries surface water from Lake Havasu on the Colorado River to a point just north of Tucson through a 538-km (336-mile) system of tunnels, pumping plants, and pipelines where it is recharged into the aquifer.

▶ Close the **Precipitation Patterns** view and open the **Supply and Demand** view, which shows Arizona counties, the CAP canal system, and selected cities that receive CAP allotments.

▸ Activate the **Arizona Cities** theme.

▸ Click the Open Theme Table button 🏢 to view the **Arizona Cities** theme table.

The **Arizona Cities** theme table includes the annual CAP allotment (in m³) of selected Arizona cities.

17. What is Tucson's annual CAP allotment, in cubic meters (m³)?

18. Using Table 2 on page 103, how does the amount of CAP water allotted to Tucson compare with the amount of precipitation that is recharged annually?

▸ Close the **Arizona Cities** theme table.

How is Tucson's water supply used?

Like a bank account, maintaining a water balance requires that the water removed from the aquifer is replaced. It is important, then, to keep track of who is using water and how much is being used. In this section you will look at how water is used in Pima County, of which Tucson is the largest city.

Water is used for many different purposes beyond drinking and typical household uses.

19. List examples of how water is used by each of the following water use sectors.

a) Public Water Supply

b) Mining

c) Agriculture

d) Industrial

What is Public Water Supply?

Public water supply includes water used in homes and businesses, as well as public use in parks and recreational facilities.

20. Speculate on the percentage of ground water allotted for use in the public water supply, mining, agriculture, and industry? Draw a pie diagram showing your estimates. Be sure to label each segment.

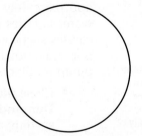

▸ Turn off the **CAP Canals** theme and turn on the **Water Usage** theme.

This theme displays pie diagrams that show the breakdown of water usage for each county in the state.

21. Draw and label the actual pie diagram for ground water usage in Pima County.

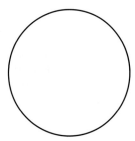

22. How did your estimate compare to the actual diagram? Which part of your estimate was most in error?

Tucson's water balance sheet

What does a water balance "look" like?

A water balance is achieved any time the balance is greater than zero. Negative values in the spreadsheet graph or table indicate a water deficit. In other words, water is being used faster than it is being replaced.

As mentioned earlier, assessing Tucson's water situation is similar to balancing a checkbook. There are deposits: recharge from precipitation, treated waste water, and CAP water into the aquifer. There are also withdrawals from the aquifer: water used by the people and businesses of Tucson. The goal, like balancing a check book, is to make sure that there are enough water deposits into the aquifer to replace the amount of water withdrawn from the aquifer.

In the Tucson Water Balance Sheet, you will balance the "water check-book" to see if there are enough deposits to match the withdrawals from the aquifer based on the city's population in 2000 and projected into the year 2050.

▸ Quit ArcView and do not save your changes.

▸ Locate the **ground_water.xls** file in the **EWR_Unit_4** folder and open it in Microsoft Excel.

This spreadsheet allows you to adjust the rate of population growth, the rate of increase in water usage, and the amount of average precipitation per year. By adjusting these variables, you can find combinations of these factors that will result in maintaining a water balance.

Examine the water balance graph.

23. If growth and water usage (including the use of CAP water) in Tucson continues at the current rate, in what year will Tucson exceed its water balance (as indicated by negative values)?

▶ Practice adjusting each of the variables to get a feel for how each will affect the water balance.

▶ Set each variable back to its **average** value when you are done.

24. If you increase the population growth rate from the current 2.4% to a rate of 5%, when will Tucson's water use exceed the supply...

   a. ...with CAP water?

   b. ...without CAP water?

25. With a 0.1% rate of increase in water usage and an average annual precipitation of 30 cm, what is the maximum population growth rate that Tucson can support (with CAP water) and still achieve a water balance in 2050?

26. With a 0.3% increase in annual water usage, what is the maximum population growth rate Tucson can support (with CAP water) to achieve a water balance in 2050? (Hint: decrease the population growth rate by 0.1% at a time.)

Scientists have predicted a trend of hotter summers and drier winters in the future.

27. If Tucson's annual precipitation drops to 21 cm per year, what combination of population growth rate and rate of increase in water usage would the city need to maintain a water balance with CAP in 2050? Find 3 combinations and enter them in Table 3 below.

Table 3—Maintaining Tucson's water balance with decreased precipitation

Combination	Population Growth Rate (%)	Water Use Annual Increase (%)	Annual Precipitation (cm)
1			21
2			21
3			21

28. You have investigated the "water checkbook" and have seen how the withdrawals balance the deposits. List three ways in which Tucson could achieve a water balance by 2050.

29. Which of the three ways do you think is the most workable? Why?

▶ Close the Excel spreadsheet. Do not save your work.

Activity 4.3

The Tucson Basin aquifer

Aquifer basics

An aquifer is a formation of rock or sediment that is saturated with water. The geology of an aquifer and its overlying sediments is an important part of understanding water infiltration and recharge.

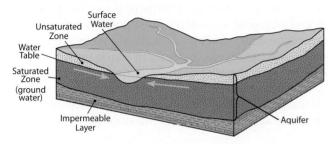

Figure 1. This diagram illustrates the structure of a typical aquifer. Rivers and lakes form where the surface drops below the water table.

The ability of water to infiltrate, or percolate down through the sediments to reach the aquifer is dictated by **permeability**. If an aquifer is permeable, water can move into it from overlying sediments. The ability of water to be stored in an aquifer is determined by **porosity**. If an aquifer is porous, it possesses empty spaces capable of retaining water. As shown in Figure 2, the quality of an aquifer is determined by both the permeability and porosity of the rock formation. Valuable aquifers are both permeable and porous, the empty spaces are connected and allow water to flow from one area to the next.

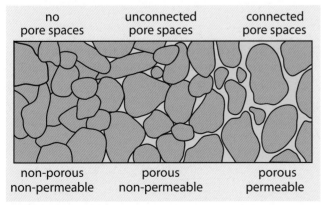

Figure 2. High porosity and permeability are characteristic properties of productive aquifers.

The Basin and Range aquifers

Tucson is located in the Basin and Range Province, a geographic region that extends from Southern Idaho into Sonora Mexico, and from Eastern California to Central Utah. This region is characterized by mountain ranges that run north-south and are separated by low basins. Viewed from the air, the Basin and Range region has been described as looking like "an army of caterpillars marching toward Mexico."

Figure 3. This map shows the location of Tucson, Arizona and the extent of the Basin and Range aquifers.

Sediments eroded from the surrounding mountains fill the basins, and can be from a few hundred to more than 3,000 meters thick. Over millions of years, water has infiltrated these sediment-filled basins, forming the Basin and Range aquifers. In these aquifers, the depth of the **water table**—the upper boundary of the aquifer's saturated zone—varies considerably, ranging from the surface (in flowing rivers and streams) to nearly 400 meters below the surface. The Basin and Range aquifer system includes more than 72 independent hydrologic basins covering over 200,000 square kilometers. These aquifers have played a critical role in the growth and development of the southwestern U.S.

The aquifer in the Tucson Basin is composed primarily of sedimentary rock including sand and clay. The cross-sectional diagram of the basin below shows the composition of the sedimentary rock that fills the basin and the historic and current location of the water table. In 1940, the water table was located in the sand layer, much closer to the surface. Since then, it has dropped more than 50 m.

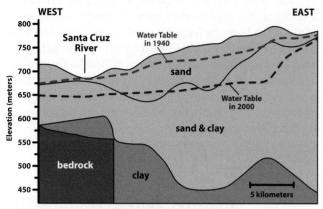

Figure 3. This cross-section of the Tucson Basin aquifer shows changes in the water table from 1940 to 2000.

Recharging the aquifer

Water removed from an aquifer is replaced or recharged primarily by precipitation. However, the majority of precipitation contributes to other parts of the hydrologic cycle and does not recharge the aquifer. Some water becomes runoff and flows into drainages. Since the water table is so low in Tucson, streams and rivers that form the drainage network are dry except during periods of heavy precipitation. Water is also lost through evaporation from the soil and absorption and transpiration by plants. The remaining 5% of the precipitation falling in the Tucson Basin eventually rejoins the aquifer, although the process can take quite a long time. University of Arizona scientists studying recharge of the Tucson Basin aquifer have found that precipitation from the 1960s still has not infiltrated down to the main aquifer.

Keeping up with demand

It is important to balance the removal of water from the aquifer and the rate at which the aquifer is being naturally or artificially recharged. There are severe environmental consequences to withdrawing more water than is recharged. When the water table declines, the aquifer may be compressed by the weight of the overlying sediments.

Compaction closes the empty spaces that previously held water, reducing the porosity and permeability of the aquifer, thus slowing infiltration and reducing the aquifer's storage capacity. Once damaged, the aquifer's porosity and permeability cannot be restored.

Compaction also results in **subsidence**, the lowering of the ground surface. Subsidence can cause considerable damage to sewer, water, and gas pipes as well as buildings and roads. Sewer water pipes rely on gravity to maintain the flow so a small change in slope due to subsidence can result in backflow—a serious problem! Intensive ground water pumping can also drive up water prices and decrease water quality. As existing wells are deepened and new wells are built to reach the sinking water table, the cost of these upgrades are passed on to the consumer. Water retrieved from deeper in the Earth is also likely to be of lower quality because salinity increases with depth.

Questions

1. What type of rock is associated with the Tucson Basin Aquifer?

2. What is the principal source of recharge in the Basin and Range aquifers? In the Tucson Basin, what percentage of this source infiltrates to recharge the aquifer?

3. What problems can result from removing more ground water from an aquifer than is replaced by recharge?

Activity 4.4

Impacts of ground water pumping

Imagine coming home and finding cracks in your walls or worse, that your house is "sinking." You and your home would be one of the many victims of *subsidence* across the country, a problem resulting from the over-pumping of ground water. In this section, you will determine the extent and severity of subsidence in the Tucson Basin.

The physical effects of subsidence

Radar Interferometry

To see a 3-D model of the subsidence in the Tucson Basin, click the Media Viewer button 🎞 and choose **Interferogram Movie**. In the animation, changes in elevation have been greatly exaggerated for emphasis.

Figure 1. This simulated cross-section through the center of the Tucson Basin (dashed blue line in top image), shows recent subsidence. Vertical distances are exaggerated.

▶ Launch ArcView and locate and open the **tucson_water.apr** file.
▶ Open the **Physical Impacts of Subsidence** view.

This view shows a *radar interferogram* of the central Tucson Basin. Radar measurements were made by satellite in 1993 and again in 1997, to precisely measure the elevation of the basin floor. By subtracting the 1997 elevation data from the 1993 data, scientists determined changes in elevation that occurred between the two dates. The elevation changes are color-coded, and appear as *interference fringes*, similar to the rainbows you see on the surface of a soap bubble due to tiny differences in the thickness of the bubble wall. In the interferogram, each cycle of colored fringes—from one blue fringe to the next blue fringe, for example—equals 2.8 centimeters of elevation change. The fringes were then superimposed on a satellite photo for reference.

▶ Activate the **Cross-section** theme.
▶ Using the Hot Link tool ⚡, click on the blue cross-section line.

The window that appears shows a cross section of the Tucson Basin aquifer and the location of the water table in 1940 and 2000.

1. Describe the change in the water table elevation from 1940 to 2000. According to the cross-section, where has the greatest change in the water table elevation occurred?

The cause of subsidence

To see a simulation of the compaction process that causes subsidence, click the Media Viewer button 🎞 and choose **Compaction Movie**.

2. If the ground water withdrawal continues at the same rate as it has since 1940, draw and label your prediction of the depth of the water table in 2050 on the diagram below.

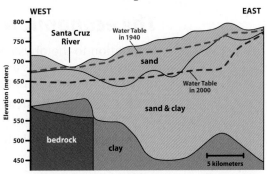

▶ Close the Hot Link window and turn off the **Cross-section** theme.

▶ Examine the **Radar Interferogram** theme.

3. How many interference fringes do you see? (Count the blue rainbow bands.)

▶ Turn on and activate the **Radar Subsidence (cm)** theme.

This theme summarizes the subsidence data shown in the radar interferogram, with each interference fringe shown as a different color.

▶ Click in each colored subsidence region using the Identify tool 🛈 to collect data about that region. Read the maximum subsidence and area from the Identify Results window, record them in the table below, and close the results window.

4. For each subsidence region, record the maximum subsidence that occurred and the area of the region in Table 1 below. Round the area to the nearest whole number.

Table 1—Subsidence in the Tucson Basin

Band	Maximum Subsidence (cm)	Area (km²)	Rate of Subsidence (cm/yr)	% of City Affected
Outer				
Middle				
Inner				

Calculating Rate of Subsidence

Rate of Subsidence (cm/yr) =
Maximum Subsidence (cm) / 4 years

5. Calculate the rate of subsidence (to the nearest tenth of a centimeter per year) for each band over the 4-year period from 1993-1997 and record it in Table 1.

6. If the city of Tucson covers an area of about 500 km², calculate the percentage of the city covered by each band (to the nearest tenth of a percent) and record it in Table 1.

Calculating % of City

Percent Area =
Area (km²) /500 km² × 100

7. If the subsidence rate remains constant, how much subsidence will occur in the center of the basin between 1997 and 2025?

▶ Close the Identify Results window.

The economic impact of subsidence

In addition to the physical impacts of over-pumping ground water, there are economic impacts as well. In this section, you will estimate the cost of repairing damage to the sewer system caused by subsidence.

▶ Close the **Physical Impacts of Subsidence** view and open the **Economic Impacts of Subsidence** view.

▶ Turn on the **Subsidence Zones (m)** theme.

The subsidence you calculated above assumed that the population and rate of water use in the Tucson Basin will remain the same. This theme shows the subsidence predicted by the US Geological Survey for the Tucson Basin between now and the year 2025, based on expected changes in population and water use.

▶ Turn on and activate the **Sewer Damage** theme.

The **Sewer Damage** theme shows the locations of main sewer lines within the predicted subsidence area. The flow in sewer lines is controlled by gravity, so even small changes in elevation could cause sewer lines to break, stop flowing, or even flow backward!

▶ Click the Open Theme Table button 🖬 to open the **Sewer Damage** theme table.

▶ Click the **Length (m)** field heading. The field name will be highlighted gray to show that it is selected.

▶ Choose **Field ▶ Statistics....**

The **Sum** reported in the statistics window is the length of sewer mains within the subsiding area, in meters.

8. What is the total length of sewer mains within the subsiding area?

▶ Close the **Sewer Damage** table.

Approximately 80% of the sewer mains will need to be replaced each time the ground in the center of the basin subsides 0.3 meters. Next, you will determine how much it might cost the city to replace these sewer mains. To simplify your calculations, assume that subsidence across the basin between now and 2025 is 1.5 meters.

9. Determine the length of sewer mains that must be replaced each time by multiplying the total length of sewer mains in the subsiding area by 80% (0.80).

10. Determine the number of times the sewer mains will need to be replaced by dividing the **amount of subsidence in 2025** by **0.3 meters**, the amount of subsidence that will require sewer main replacement.

11. Determine the total length of sewer mains that must be replaced by 2025 by multiplying the **number of replacements** (from question 10) by the **total length of sewer mains in the subsiding area** (from question 9).

12. According to a local engineering firm, the cost of replacing a sewer main is $200/meter. How much will it cost the city to replace the damaged lines over the next 25 years?

Activity 4.5

Conserving water

Tucson's water rates

Water rates increase over time for several reasons. Rates must be raised to cover the costs of operating, maintaining, and expanding the water system, and paying employee salaries. Water rates may also be increased to encourage customers to conserve water. The reasoning is that, as water rates increase, customers will decrease their water use in order to avoid higher monthly bills. In this section, you will investigate how residential water rates have changed through time in Tucson. You will also determine how much water *really* costs.

Table 1—Tucson residential water rates

Tucson Residential Water Rates (1925–2000)					
Population	**Year**	**Class**	**Price ($/Ccf)**		**w/ Inflation**
			Winter	Summer	Year 2000 $
790755 (1988)	2000	Residential			
		0–3 Ccf	$0.00	$0.00	$0.00
		4–15 Ccf	$1.62	$1.62	$1.62
		16–30 Ccf	$2.61	$2.61	$2.61
		over 30 Ccf	$3.29	$3.29	$3.29
531443 (1980)	1980	Residential			
		0–10 Ccf	$0.69	$0.69	$1.44 / $1.44
		11–20 Ccf	$0.71	$0.80	$1.48 / $1.67
		21–50 Ccf	$0.73	$0.94	$1.52 / $1.96
		over 50 Ccf	$0.78	$1.09	$1.63 / $2.28
265660 (1960)	1964	Residential			
		0–37 Ccf	$0.20	$0.20	$1.11
		over 37 Ccf	$0.18	$0.18	$1.00
141216 (1950)	1952	Residential			
		unlimited Ccf	$0.10	$0.10	$0.65
55676 (1930)	1925	Residential			
		unlimited Ccf	$0.09	$0.09	$0.88

What is $/Ccf ?

The price of water is given in the units $/Ccf which stands for the price of water per 100 cubic feet of water.

What is inflation?

Inflation is an increase in the prices of products and services over time. Inflation can also be described as the decrease in the purchasing power of money over time. By "adjusting for inflation," prices at different times can be realistically compared.

Table 1 shows a sample of water rates in Tucson over time (**Price ($/Ccf)** stands for the price of water per 100 cubic feet of water). Prior to 1978, there was no distinction between summer and winter water costs. Today, commercial, industrial, and multi-family classes pay under a two-tiered system. The **w/ Inflation** column shows the cost of water rates from the past adjusted for inflation to the year 2000.

The graphs below show the population and water rates in the Tucson metropolitan area over several decades.

Graph 1—Tucson Area Population

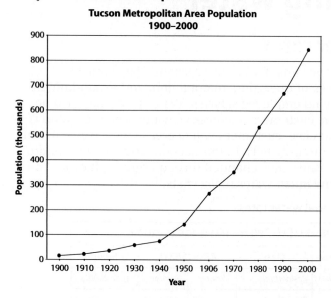

Graph 2—Tucson Area Water Rates

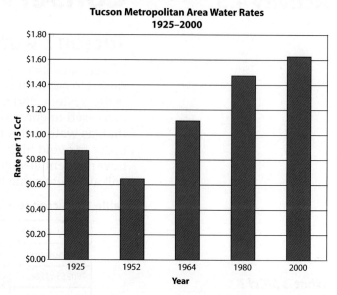

1. According to Graph 1, how large was the Tucson Metropolitan Area population in 1925? How large was the population in 2000?

2. Calculate the percent increase in population by subtracting the 1925 population from the 2000 population, divide that number by the 1925 population, and multiply by 100.

3. According to Graph 2, what was the cost of water in the Tucson Metropolitan Area in 1925, in dollars per hundred cubic feet ($/Ccf)? What was the cost in 2000?

4. Calculate the percent increase in water rates by subtracting the 1925 water rate from the 2000 rate, divide that number by the 1925 population, and multiply by 100.

5. Is the percent increase in water costs more, less, or similar to the rate of population increase?

The following table shows water usage (gallons per day) for a family of four to do certain jobs around the house.

6. Calculate the amount of water used by a family of four each month for these tasks. The first one has been done for you.

Table 2—Typical household water usage for a family of four

Job	Gallons per day	Times per month	Gallons used per month
Shower	80	30	80 x 30 = **2400**
Flush Toilet	100	30	
Brush Teeth	8	30	
Wash Dishes	15	30	
Cooking & Drinking	12	30	
Irrigation	48	30	
Kitchen Sink	5	30	
Laundry (1 load)	35	17	
		Total Gallons	

7. Convert **Total Gallons** calculated in the table to hundreds of cubic feet (Ccf) per month. (Hint: 1 Ccf = 746 gallons)

8. Using the value you calculated for Ccf per month in question 7, calculate your monthly water bill for both winter and summer during each of the years listed in Table 3 below. Use Year 2000 values (dollar values adjusted for inflation) in Table 1.

Table 3—Estimated summer monthly water bill for a family of four

Year	Ccf per month	x Winter Rate (Year 2000$)	Monthly Winter Bill	x Summer Rate (Year 2000 $)	Monthly Summer Bill
Ex. 1980	27	$1.52	$41.04	$1.96	$52.92
2000					
1980					
1964					
1952					
1925					

9. How much has the monthly water bill increased since 1925?

10. Calculate the cost per gallon of water in 2000 by dividing the amount of the summer monthly bill (from Table 3) by the **Total Gallons** you calculated in the Table 2. This new number is the approximate cost per gallon of water in 2000.

11. If bottled water costs $1.00 per gallon, how many more times expensive is bottled water compared to tap water? (Divide $1.00 by your answer to question 10.)

Imagine what it would cost to meet all of your family's water needs with bottled water!

12. How do you think doubling water rates would affect the city's population and the city's water supply? Explain.

Stop the pumping!

You have investigated subsidence and its potential impacts: damage to homes, and damage to water mains and sewer lines. In theory, the solution to many of these problems is easy: stop pumping ground water from subsiding areas. How would this affect Tucson's water supply?

▶ Launch ArcView and locate and open the **tucson_water.apr** file.

▶ Open the **Wells** view.

This view shows the major streets in the city of Tucson and the boundary of the zone where ground subsidence is occurring.

▶ Turn on and activate the **Wells** theme.

This theme shows the drinking water wells in the Tucson Active Management area.

▶ Select the red subsidence zone boundary using the pointer tool [▶]. The boundary will show rectangular handles around it when it is selected.

▶ Click the Select By Graphic button [⬚].

The drinking water wells within the subsidence zone will be highlighted yellow.

▶ Click the Open Theme Table button [▦] to open the **Wells** theme table.

▶ Read the number of selected wells on the tool bar.

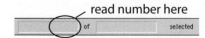

read number here

13. How many drinking water production wells are located within the central Tucson subsidence zone?

▶ Click the **Production (gal/yr)** field heading. The field name will be highlighted gray to show that it is selected.

This field gives the pumping rate, in gallons per year, for each well.

▶ Choose **Field ▶ Statistics....**

In the statistics window, the **Sum** is the total amount produced by the highlighted wells in the subsidence zone.

14. If these 46 wells within the subsiding area were removed from production, how many gallons would be lost per year?

15. How do you think the city might make up for this shortfall?

▶ Close the statistics window, the **Wells** theme table, and the **Wells** view.

Water conservation

Research in the Tucson Basin has shown that only about 5% of the precipitation that falls in the basin is recharged into the aquifer. In urban areas, a significant portion of the precipitation runs off into storm drains and sewer systems. In this section, you will calculate how much water can be harvested from a small area during a single summer storm.

▶ Open the **Summer Harvest** view.

This view shows the major streets in the city of Tucson and an aerial photo of the University of Arizona. The bold yellow line outlines the main campus area of the university.

▶ Turn on and activate the **University Buildings** theme.

This theme is a base map of the campus buildings and athletic facilities.

▶ Use the Pointer tool 🖈 to draw a box around the main campus, indicated by the bold yellow line.

The area of the region you outlined will be displayed on the status bar.

read area here

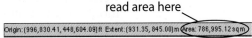

Origin: (996,830.41, 448,604.09) ft Extent: (931.35, 845.00) m Area: 786,995.12 sq m

16. What is the area of the main campus in square meters?

17. Assuming that buildings cover about 30% of the main campus area, and that the rest is open space, calculate the area of the main campus occupied by buildings (in square meters).

Watch a summer thunderstorm!

To view time-lapse movies of summer thunderstorms, click the Media Viewer button 🎬 and choose **Thunderstorm Movie 1** or **2**.

18. Following the directions below, calculate how much water could be collected from the rooftops of the buildings on campus if the storm produced 2.5 cm (1 inch) of rain.
    a) Convert centimeters to meters of rain (100 cm = 1 m).

    b) Calculate the volume of rain on the rooftops, in cubic meters. (Hint: volume = area of rooftops (m^2) calculated in question 17 × amount of water (m) calculated in question 18a).

    c) Convert the volume of water collected from the rooftops to gallons (1 m^3 = 264.2 gallons).

19. How would you collect water from the rooftops of the buildings in the main campus area?

20. What kinds of things can this harvested water be used for?

21. What percentage of the average annual well production in the subsidence zone is the volume of water harvested from campus rooftops from 2.5 cm of rainfall? (Hint: answer 18c ÷ answer 14 × 100)

22. Given that an average-sized house is 149 m^2, how much water could be collected from the roof of an average house in the same storm? Report your answer in both cubic meters and gallons.

23. Referring to Table 2, what percentage of the average family's monthly water allotment for irrigation could be met using the water harvested from the rooftop in this storm?

24. Do you think this is a valid technique for conserving water in Tucson? What issues might complicate this type of conservation?

Activity 4.6

The voice of conservation

Voicing your ideas on conservation

Now that you have examined Tucson's water situation, and the environmental, physical, economical, and political issues that are part of the situation, you may have developed some of your own solutions to the water supply problem. This section provides you with an opportunity to express your ideas and develop a valid conservation plan for the city of Tucson, using the knowledge you gained in this activity.

Project

You have just been hired by the local water department of Tucson as a community water conservation liaison. Your job is to communicate with citizens ways to conserve water and the importance of doing so. Write a letter that will be mailed to all residents explaining why they should care about conserving water and what things they can do to save water. In this letter, you should describe three different action plans for residents to help decrease groundwater use in Tucson. Give specific facts and details about your plans and about what will happen if Tucsonans do not conserve and if they do. You might want to refer to values you calculated in Activity 4.2 (the Tucson water balance sheet) and Activity 4.5.